AI and IoT in the EduTech Era

Artificial intelligence (AI) and the Internet of Things (IoT) are redefining the educational landscape across schools, universities, and training institutions worldwide. This book explores how these technologies are reshaping teaching, learning, assessment, governance, and institutional strategy in the evolving EduTech era.

Bringing together international scholars and practitioners, this volume examines the integration of AI and IoT from multiple perspectives: pedagogical, institutional, strategic, ethical, and environmental. It addresses resistance to technological change, the development of smart classrooms, prompt literacy for educators, AI-ready school models, and policy-to-practice transitions in higher education. The book also explores teacher identity transformation through AI co-teaching tools, equity challenges in global EdTech ecosystems, and empirical insights such as evidence from British universities. Furthermore, it analyzes smart diagnostics, personalized learning systems, sustainability concerns related to AI and IoT infrastructures, and the broader connection between digital transformation and education. Through conceptual frameworks, case studies, and policy analysis, the book provides a comprehensive roadmap for responsible and effective AI-driven educational innovation.

This book is intended for researchers, policymakers, school leaders, higher education administrators, EdTech professionals, and graduate students in education, digital transformation, and innovation management. It will also benefit practitioners seeking strategic guidance for implementing AI and IoT solutions in educational environments.

Emerging Trends in Science, Innovation and Business
Series Editor: Saad Motahhir

Instructional Design in E-Learning: Principles, Trends, and Opportunities
Edited by Soufiane Ouariach, Fatima Zahra Ouariach, Inam Ullah Khan, Firyal Latrache, Najwa Jbira, and Veerpratap Meena

For more information about this series, please visit: www.routledge.com/Emerging-Trends-in-Science-Innovation-and-Business/book-series/ETSIB

AI and IoT in the EduTech Era

The Future of Education

Edited by

Yahya Fikri and Adnène Arbi

CRC Press
Taylor & Francis Group
Boca Raton London New York

CRC Press is an imprint of the
Taylor & Francis Group, an **informa** business

Designed cover image: Shutterstock

First edition published 2027
by CRC Press
4 Park Square, Milton Park, Abingdon, Oxon, OX14 4RN

and by CRC Press
2385 NW Executive Center Drive, Suite 320, Boca Raton FL 33431

CRC Press is an imprint of Taylor & Francis Group, LLC

ISBN: 978-1-041-29061-2 (hbk)
ISBN: 978-1-041-29064-3 (pbk)
ISBN: 978-1-003-76785-5 (ebk)

DOI: 10.1201/9781003767855

Typeset in Times
by KnowledgeWorks Global Ltd.

Contents

About the Editors ... vii
Contributors ... ix

Chapter 1 Getting Past Opposition to Change: Strategies for AI and IoT Integration in Education ... 1
Yahya Fikri, Rocky Kumar, and Arshi Naim

Chapter 2 Beyond the Hype: Artificial Intelligence in Education ... 21
Yahya Fikri, Rocky Kumar, Bhanuprakash Madupati, Abdelhamid Nechad, Mohamed Rhalma, and Redouane Kaiss

Chapter 3 The Smart Classroom and IoT ... 37
Revati Ramrao Rautrao

Chapter 4 Prompt Literacy for Educators: A Framework for Effective AI Interaction in Lesson Planning ... 53
Adil El jouali and Wiame El amri

Chapter 5 Creating an AI-Ready School ... 61
Husni Muhamad Rifqi, Tajudeen Alaburo Abdulsalam, and Yahya Fikri

Chapter 6 From Policy to Practice: Exploring AI Integration in Higher Education ... 71
Meet Purohit and Bhavna Sharma

Chapter 7 AI and IoT-Enabled Smart Classrooms: Redefining Teaching–Learning Processes in Higher Education ... 78
Shyamasundar Tripathy, José G. Vargas-Hernández, and Islam Elgammal

Chapter 8 Global Perspectives on EdTech Equity ... 89
Amine Hmid, Achy Abdelghafour, Imane El Moumni, Mehdi Lechquar, and Redouane Kaiss

Chapter 9 How GenAI Creates Absorptive Capacity for Business Strategy Educators: A Survey from British Universities ... 99
Mostafa Mohamad, Sean Chung, and Adelina Emini

Chapter 10 Act for an AI School: Smart Diagnostics and Personal Support ... 117
Nadia Elassassi, Said Machwate, Rachid Bendaoud, and Said El Hasri

Chapter 11 AI-Enabled Education: Implications for Teaching, Learning, and Assessment 125

Kamini Bhasin

Chapter 12 Environmental Sustainability of IoT and AI Deployments in Education: Pathways, Challenges, and Green Solutions 140

Ajay Kumar, Yahya Fikri, Sanjeev Kumar, Kuldeep Singh, and Redouane Kaiss

Chapter 13 AI and IoT-Enabled Smart Classrooms: Redefining Teaching–Learning Processes in Higher Education 150

Shyamasundar Tripathy, Sorie Saidu Kamara, B. Vamsi Krishna, V. S. Prasad Kandi, and Umesh Kumar Shukla

Chapter 14 From Authority to Advisor: How AI Co-Teaching Tools Redefine Teacher Identity 161

Shyamasundar Tripathy, José G. Vargas-Hernández, and Islam Elgammal

Index 181

About the Editors

Adnène Arbi is an Associate Professor affiliated with the University of Carthage in Tunisia, where he has contributed significantly to the fields of mathematics, AI, and engineering. His tenure at the university spans over a decade, during which he has held various positions, including roles at the Tunisia Polytechnic School and the National Institute of Applied Sciences and Technology. His research interests primarily focus on neural networks, dynamical systems, and mathematical modeling, with a particular emphasis on the stability and controllability of complex systems. Arbi's recent publications reflect his commitment to advancing knowledge in applied mathematics, particularly through innovative approaches such as wavelet neural networks and the dynamics of nonlinear systems. His work has been published in reputable journals, contributing to the understanding of mathematical phenomena in both theoretical and practical contexts. In addition to his academic pursuits, Arbi has collaborated with various prestigious institutions globally, including the University of the Witwatersrand, University of Southeast, and Imperial College London, showcasing his dedication to fostering international research partnerships. His contributions to the academic community continue to inspire emerging scholars and practitioners in the field. Actually, Dr. Arbi served as editor of various journals. Besides, he is the author of more than 50 scientific peer-reviewed papers in international journals and conference proceedings, and he served as a reviewer for many international conferences and workshops.

Dr. Yahya Fikri is a PhD researcher in *Economic and Management Sciences*, specializing in *economics, econometrics, and statistics*, with a strong focus on applying advanced quantitative and qualitative methods to complex economic challenges. His expertise spans *time series modeling, data analysis, and statistical inference*, areas in which he has consistently advanced both theory and application.

Over the course of his academic career, Dr. Fikri has been a distinguished academic who has published extensively in *peer-reviewed journals*, where his contributions have significantly shaped the development and refinement of modern econometric techniques. His scholarly influence is complemented by active participation in *international conferences and seminars*, where he fosters interdisciplinary collaboration and knowledge exchange across economics, environmental science, and policy domains. And, Dr. Fikri is an associate editor and reviewer for some journals and book chapters.

Dr. Fikri has emerged as a leading voice in examining the *nexus of education, poverty, economic growth, and sustainable development.* By leveraging applied statistics and econometric modeling, his research provides critical insights into the *socio-economic impacts of education on poverty, economic growth, and long-term sustainability.*

Beyond academic contributions, his work carries profound societal relevance. By generating *evidence-based models and policy recommendations*, Dr. Fikri supports strategies aimed at enhancing *economic growth, AI in education, and environmental sustainability.* His research exemplifies how *rigorous data-driven analysis* can bridge the gap between economic theory and practical solutions to some of the most urgent global challenges.

With his blend of methodological precision, interdisciplinary engagement, and commitment to sustainability, Dr. Fikri continues to shape the future of *econometric-driven policy and sustainable development research.*

Contributors

Tajudeen Alaburo Abdulsalam
Sheldon B. Lubar College of Business
University of Wisconsin-Milwaukee
Milwaukee, WI, USA

Abdelghafour Achy
Education, Languages, and Cultures, ENS
Moulay Ismail University
Meknes, Morocco

Rachid Bendaoud
Faculty of Science Semlalia
Cadi Ayyad University
Marrakech, Morocco

Kamini Bhasin
Jaypee University of Engineering & Technology
Guna, India

Bhargab Pratim Bora
B.H. College
Bhattadev University
Howly, India

Sean Chung
Salford Business School
University of Salford
Manchester, United Kingdom

Wiame El amri
Faculty of Legal, Economic and Social Sciences, Ait Melloul
Ibn Zohr University
Agadir, Morocco

Said El-Hasri
Faculty of Science Technology
Cadi Ayyad University
Marrakech, Morocco

Adil El jouali
Faculty of Legal, Economic and Social Sciences, Ait Melloul
Ibn Zohr University
Agadir, Morocco

Imane El moumni
Hassan first University
Settat, Morocco

Islam Elgammal
Collage of Business
University of Business and Technology
Jeddah, Saudi Arabia

Nadia Elassassi
Faculty of Science Semlalia
Cadi Ayyad University
Marrakech, Morocco

Adelina Emini
Marketing Department
University of Prishtina
Pristina, Kosovo

Yahya Fikri
National School of Business and Management
Abdelmalek Essaadi University
Tangier, Morocco

Amine Hamid
Faculty of legal economic and social sciences (Ain Sebâa)
Hassan II University
Casablanca, Morocco

Redouane Kaiss
Faculty of Economics and Management
Hassan 1st University
Settat, Morocco

Sorie Saidu Kamara
Campus Freetown Polytechnic, Jui Junction
Sierra Leone, West Africa

V.S. Prasad Kandi
KL Business School
Koneru Lakshmaiah Education Foundation
Guntur, India

B. Vamsi Krishna
KL Business School
Koneru Lakshmaiah Education Foundation
Guntur, India

Ajay Kumar
Chandigarh College of Hospitality
Chandigarh, India

Sanjeev Kumar
Lovely Professional University
Chandigarh, India

Mehdi Lechguar
Hassan 2 University
Casablanca, Morocco

Said Machwate
Higher National School
Cadi Ayyad University
Marrakech, Morocco

Bhanuprakash Madupati
Department of Corrections
Texas, USA

Mostafa Mohamad
Zayed University
Dubai, United Arab Emirates

Arshi Naim
London Institute of Sustainable Development
London, United Kingdom

Abdelhamid Nechad
School of business and Management, ESCA
Casablanca, Morocco

Meet Purohit
School of Commerce, Finance & Accountancy
Christ University
Delhi, India

Revati Ramrao Rautrao
Patil B-School
Pune, India

Mohamed Rhalma
National School of Business and Management
Abdelmalek Essaadi University
Tangier, Morocco

Husni Muhamad Rifqi
KFUPM Business School
King Fahd University of Petroleum & Minerals
Dhahran, Saudi Arabia

Bhavna Sharma
School of Commerce, Finance & Accountancy
Christ University
Delhi, India

Smita Sharma
MNC Balika Mahavidyalaya
Gauhati University
Nalbari, India

Kuldeep Singh
Amity University
Haryana, India

Umesh Kumar Shukla
KL Business School
Koneru Lakshmaiah Education Foundation
Guntur, India

Shyamasundar Tripathy
KL Business School, Koneru Lakshmaiah
Education Foundation
Guntur, India

José G. Vargas-Hernández
Professor, Financial Literacy Research Center
Universidad Cristóbal Colón
Veracruz, Mexico

1 Getting Past Opposition to Change

Strategies for AI and IoT Integration in Education

Yahya Fikri, Rocky Kumar, and Arshi Naim

1.1 INTRODUCTION

The global education ecosystem is on the brink of a drastic change and it is due to the swirl of developments and innovations in artificial intelligence (AI) and the Internet of Things (IoT). From intelligent tutoring systems and adaptive learning platforms to smart classrooms and predictive analytics, new technologies are redefining the delivery, assessment and management of knowledge. Governments, along with universities and private educational institutions, are making a growing investment in digital infrastructure in order to modernize schooling environments, matching the requirements of the Fourth Industrial Revolution. However, change in technology and education isn't just a change in technology – it is a socio-organizational transition. While there are unprecedented possibilities to personalize, make things more efficient and data-driven through integrating AI and IoT, there are often institutional resistance, cultural hesitation, ethical concerns and structural barriers. Educational institutions, which have been socialized in traditional pedagogical practices, may often challenge themselves to comply with disruptive technological paradigms. This tension between innovation and resistance gives rise to a critical need for some kind of structured change management strategies. This chapter discusses the root causes of regimes of opposition to AI and IoT integration in education as well as strategic framework for moving educational institutions from regime of resistance to readiness. By drawing from a combination of change management theory, governance models and global best practices, the chapter offers a road map to sustainable and ethical technological change in education (Akgun & Greenhow, 2022).

1.2 THE DIGITAL TRANSFORMATION IMPERATIVE IN EDUCATION

Digital transformation in education is no longer even an option; it is a matter of operations and strategic requirements. The growing need for remote learning, hybrid models of education and data-driven educational decisions has accelerated institutional digitization globally. AI-powered analytics are now helping to provide capabilities of identifying learning gaps, performance prediction of students and personalization of curriculum pathways. Simultaneously, various IoT-enabled devices like smart boards, environmental sensors, wearable devices and attendance tracking systems along with other enhance the application of classroom automation and campus management. As a result, educational competitiveness is more and more dependent on technology adaptability. Institutions that do not modernize fall behind in defining a new skills gap between academic preparation and workforce expectations. Moreover, with the increasing presence of digital literacy, computational thinking and AI awareness being overwhelmingly defined within educational policy around the world, there is a newly pressing need to build the capacity for these core competences as a way to encourage future-ready students. Yet, digital transformation goes beyond hardware installation or software deployment. It requires the rethinking pedagogy models, methods of assessment,

DOI: 10.1201/9781003767855-1

governance structures and mechanisms for teachers training. The imperative is not merely the digital adoption, but strategic digital integration elaborated by institutional vision and protection of ethical safeguards (Alhumaid et al., 2020).

1.3 AI AND IoT AS CATALYSTS OF SYSTEMIC CHANGE

AI and IoT technologies are no longer used as a set of additional tools but in a systemic and transformative way in the educational ecosystem today. AI-driven platforms enable dynamic and adaptive learning environments where the content of instruction can be dynamically adjusted to student performance, learning path and cognitive patterns of students. Machine learning algorithms allow early detection of learning gaps and guide specific learning intervention strategies, as well as optimize administrative tasks, such as grading and performance tracking. Simultaneously, IoT infrastructures connect devices, sensors and digital platforms creating a data collection system in real-time, considering automated attendance management, environment optimization and coordination of operations all around campus. The convergence between AI and IoT thus creates a source of intelligent educational ecosystems, which have the potential to foster not only the effectiveness of the pedagogy but also the efficiency of the institution. However, these transformations are not limited to the classroom innovation, but also impact on the governance structures, resource allocation strategies, compliance mechanisms and policy planning for the future years. In the process, AI and IoT disrupt institutional hierarchies and decision-making processes. Yet, transformative change of systems comes to disrupt established routines and professional identities. AI is raising questions for educators about pedagogical independence issues, concern from administrators about financial stability and regulatory compliance programme and students/parents with issues related to privacy and algorithmic fairness. Therefore, AI and IoT work in parallel as enablers of progress and catalysts of resistance (Bender et al., 2021).

1.4 THE PERSISTENCE OF RESISTANCE IN EDUCATIONAL INSTITUTIONS

Resistance to change is a well-established phenomenon of organization, especially those institutions that display features of tradition, layered governance and regulatory accountability – features often applied to educational systems. The persistence of resistance results from linked psychological and structural factors, ethical issues, economic considerations and culture. Educators could be anxious about job security, technological self-efficacy or reduced professional autonomy. Institutional inertia, being landed within bureaucratic policy decision-making processes and academic rigidity, can act as a form of resistance that slows down adaptation and increases scepticism. Data privacy, algorithmic bias and surveillance are in the context of ethical and legal uncertainties that make hesitation stronger. Financial constraints linked to investment in infrastructure, system maintenance and training programmes pose additional barriers. Cultural felicity of the conventional pedagogical approaches could also reinforce scepticism to automation. Unlike private enterprises, educational institutions exist within public framework of accountability and amicability systems as well as social values systems, where a technological change is often phenetamines following a normative perspective. As such, digital transformation opposition is often framed as the defence of educational integrity and not simply a dig against innovation. Many stakeholders, such as patients, have had negative experiences with medical methods that involve waiting. Recognizing resistance as a natural and informative phase of transformation enables institutions to make sense of stakeholder concerns. With the benefit of participatory governance, clear communication and stages of implementation, resistance can become a catalyst for more ethically grounded and institutionally resilient digital reform (Bond et al., 2021).

1.5 OBJECTIVES AND STRUCTURE OF THE CHAPTER

This chapter seeks to give a comprehensive framework for overcoming the opposition to the integration of AI and IoT in education.

The specific objectives are:

1. To analyse the multidimensional causes of the resistance to technological change.
2. To test the established theories of change and technology adoption in the educational context.
3. To measure the stakeholder views and dynamics of trust.
4. To identify ethical, legal and governance issues.
5. To suggest strategic scales and phased roadmap for sustainable AI-IoT implementation up to 2035.

This chapter is organized in thematic sections. Following this introduction, the next section attempts to examine the categories and drivers of resistance. The follow-up parts cover the theoretical basis, stakeholder interactions, governance issues and strategic frameworks. Case studies and a proposed unified change management model are included before the conclusion with policy recommendations and future directions. Combining the synthesis of theory, empirical evidence and future-oriented strategies, the chapter is part of a conversation about digital transformation in education and provides institutional guidance on coping with technological change (Bond et al., 2024).

1.6 UNDERSTANDING OPPOSITION TO CHANGE IN AI AND IoT ADOPTION

The incorporation of AI and IoT technologies in the educational environment entails a transformative change in the way teaching and learning is carried out as well as the way institutions are governed. However, despite good policy advocacy coupled with technology being advanced, the push back against change remains a persistent challenge. Resistance is not usually rooted in one or other factor but instead arises from a mesh of psychological, organizational, technological, ethical and socio-cultural factors. Understanding these dimensions is important to the design of effective change management strategies. Instead of seeing resistance as obstruction, institutions must seek to interpret it as the signal of uncertainty, trust deficits or structural misalignment. This section classifies and discusses in a critical way the major forms of opposition that occur during AI and IoT adoption in educational settings (Cabrera et al., 2025).

1.6.1 Psychological Resistance Among Educators

Educators, in particular, are at the forefront of the successful implementation of AI-driven and IoT-enabled educational systems. However, psychological resistance often arises because of perceived threats to professional identity, autonomy and pedagogical authority.

One issue that directed concerns is fear of job displacement. In certain ways, the grading, content recommendation and performance monitoring systems that are automated may cause some anxiety about the role of teachers in the future. Although AI is meant to augment, instead of replace, human instruction, the connotation of replacement causes opposition.

Another major factor is technological self-efficacy. Many educators, especially those who were taught in the traditional pedagogy, may not be comfortable with the idea of leveraging advanced analytics platforms, AI dashboards or IoT-enabled classroom devices. The learning curve when a new system is used can lead to cognitive overload and stress. In addition to that, a lack or absence of pedagogical control is often specified. AI-driven adaptive systems might recommend content changes or learning interventions that can challenge the teaching methods that teachers have established. This perceived decline in professional autonomy is part of the problem of scepticism.

Psychological resistance is often exacerbated by inadequate training, limited institutional support and poor communication. (Cannarsa, 2021).

1.6.2 Institutional and Organizational Barriers

Educational institutions of learning readily work in hierarchical governance structures and inveterate administrative frameworks. Organizational inertia can be a serious slowdown to the rate of adoption of AI and IoT.

One significant obstacle is bureaucracy in decision-making. Procurement approvals, regime compliance reviews and inter-departmental coordination can cause delays in technological deployment. Unlike private enterprises, public educational institutions need to deal with frameworks of accountability and policy constraints. Lack of strategic vision is also seen as a cause of resistance. Institutions that deploy AI or IoT with little to no road map may face broken implementation, resulting in incompatibility between systems and inefficiencies in operations. In such cases, it is not technology but poor governance planning which provokes the opposition. Furthermore, the allocation of resources between different departments and administrative units can cause tension among academic departments and administration. Budget redistribution in favour of digital infrastructure potentially made feel a commitment to cut funding for traditional academic priorities.

Organizational culture plays a very critical role. Institutions with well-established teaching traditions can be resistant to disruptive innovation. Change initiatives that are not about furthering institutional mission and values are more likely to encounter sustained opposition (Chan & Colloton, 2024).

1.6.3 Technological Infrastructure Challenges

The successful integration of AI and IoT systems necessitates sound digital infrastructure such as high-speed internet connectivity, safe data storage, interoperable software platforms and good hardware maintenance systems. Many institutions, especially in the developing world, have digital infrastructure gaps. Limited bandwidth, antiquated computing power and inconsistent power supply make proper deployment of smart technologies challenging. Another challenge is the interoperability of the systems. AI platforms, learning management systems, IoT sensors and administrative software are required to work together. Incompatibility with the legacy systems to the newly introduced technologies can cause operational disruptions and lack of trust in digital transformation initiatives.

Cybersecurity vulnerabilities make adoption that much more difficult. IoT devices provide more points of access to a network and fears exist about breaches of data, attack through ransomware and manipulation of the system. If institutions do not have cybersecurity protocols, then stakeholders might oppose the integration of technologies due to perceived risk. In addition, maintenance cost and ongoing technical support burden for maintenance also exist. For preventing institutions from being reluctant to undertake more long-term digital transformation projects, some kind of sustainable funding models can be established (Chassignol et al., 2018).

1.6.4 Ethical and Privacy Concerns

AI and IoT systems rely heavily on the collection of data and processing by algorithms. In the case of educational use, this usually means sensitive student information, behaviour tracking, performance analytics and biometric monitoring. One key issue is that of student data privacy. IoT-enabled attendance systems, smart surveillance cameras, wearable learning devices, etc., can possibly produce constant stream of personal data. Without transparent data governance policies in place, stakeholders are afraid of the misuse or unauthorized access.

Algorithmic bias is another ethical issue. AI systems that are trained with the dataset of biases may lead to discriminatory results in terms of grading, admissions prediction or behavioural assessment. Such risks destroy trust and lead to questions of fairness and accountability. The issue of transparency and explainability is also important. Black-box AI models lack clear justification for their recommendations or decisions, which greatly restricts educators to challenge or interpret the outcomes. The regulatory uncertainty causes these concerns to become even worse. Institutions

TABLE 1.1
Categories of Resistance to AI and IoT in Education

Category of Resistance	Primary Source	Key Stakeholders	Core Concerns	Impact Level
Psychological resistance	Individual-level factors	Teachers, staff	Job security, autonomy, skill gaps	High
Organizational barriers	Institutional structure	Administrators, departments	Bureaucracy, strategic misalignment	High
Infrastructure limitations	Technical ecosystem	IT teams, management	Connectivity, interoperability, maintenance cost	Medium-high
Ethical and privacy concerns	Data governance	Students, parents, regulators	Surveillance, bias, transparency	High
Socio-economic and cultural resistance	Social environment	Communities, families	Digital divide, cultural values	Medium

have to come to grips with changing data protection legislation and AI governance structures. Inconsistencies in the policy environments can result in compliance anxiety and a lack of adoption (Dwivedi et al., 2021).

1.6.5 Socio-Economic and Cultural Resistance

Technological change does not happen in a social vacuum. Socio-economic differences play a huge role in the adoption of AI and IoT in learning. The digital divide continues to be a very significant obstacle. Students from economically disadvantaged backgrounds may not have access to a reliable internet connection, smart devices and supportive digital environments at home. Integrating advanced technologies without dealing with equity issues will only serve to widen educational inequality. Cultural attitudes towards technology are also a factor in resistance to technology. In communities where traditional face-to-face instruction has a deep cultural value, automated systems may be thought of as impersonal and culture disruptive to educational values. Parental concerns around screen time as well as surveillance and relying too much on automation can increase opposition even more. Public perception plays a strong role in the internal decisions of such institutions, particularly in publicly supported education systems. There is also a chance of scepticism towards the AI-driven initiatives by both the government and the technology corporations as trust deficit may arise. Without the transparency in reaching stakeholders, socio-cultural resistance can remain even though technological benefits are apparent (Floridi & Cowls, 2022).

The multidimensionality of resistance emphasizes a need for the categorization of barriers in the psychological, institutional, ethical and socio-cultural domains. A structured classification of such types of resistance is presented in Table 1.1.

The interconnected layers of resistance at the individual, institutional and policy levels are presented in a conceptual manner in Figure 1.1, showing how the combined effects of multi-level barriers shape the adoption of AI-IoT in education.

1.7 THEORETICAL FOUNDATIONS OF CHANGE MANAGEMENT

The integration of AI and IoT technologies in educational systems should be understood through the established change management and technology adoption theories. Educational institutions are complex socio-technical environments where transformation includes not only modification of infrastructure but also changes in behaviour, restructuring of governance and cultural recalibration. Theoretical bases thus offer analytical perspectives for understanding resistance as well as structured approaches to intervention (Gunasekara et al., 2025).

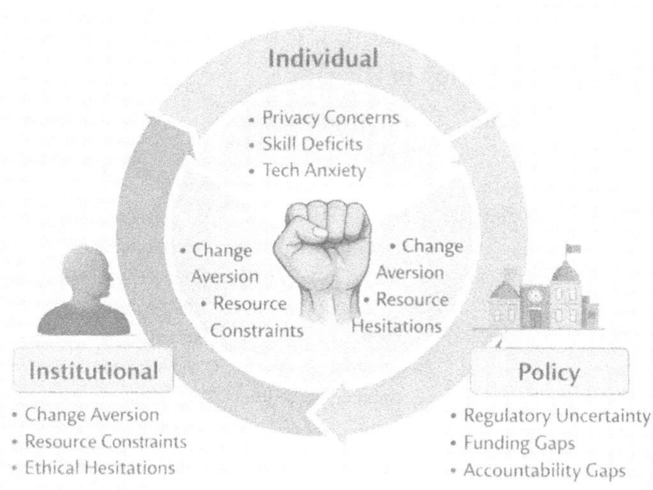

FIGURE 1.1 Multi-level resistance framework (Individual-Institutional-Policy Model).

1.7.1 Lewin's Three-Stage Model

Kurt Lewin's Three-Stage Model when it comes to change, conceptualizes the change as a progression through unfreezing, changing and refreezing. In light of AI-IoT integration, unfreezing refers to preparing educators and administration to challenge old learning norms and become alert to the need of digital transformation. The changing phase involves the implementation of AI-based instruction systems and IoT enabled infrastructure with assistance of professional development and guide from institution. Refreezing stabilizes these practices through the alignment of policies and performance evaluation mechanisms and the embedding of cultural practices. Although the model is clear and structurally simple, the linear design may require modification to accommodate the iterative change of digital technologies.

1.7.2 Kotter's Eight-Step Change Model

Kotter's Eight-Step Model Adapts Leadership Driven Transformation, Cultural Anchoring. By building urgency, creating guiding coalitions, articulating strategic vision and creating wins in the short term, institutions can begin to institutionalize AI-IoT systems into the organizational culture over time. This model is especially valuable in education where stakeholder alignment and the commitment of leadership over the long term decide the success of adopting it. However, its implementation requires long-term strategic coherence and resource allocation (Holmes & Porayska-Pomsta, 2023).

1.7.3 Technology Acceptance Model (TAM)

The Technology Acceptance Model changes the level from the organizational perspective to the individual one, by suggesting that people's perceived usefulness and ease of use have a significant effect on their decision to use the technology. To put it in the context of education systems, a teacher's decision-making journey on AI-based systems is more inclined to follow the direction of AI systems when saw as instructional useful with manageable complexity. TAM has a good track record in explaining psychological resistance but fails to cover the institutional or the regulatory dimensions fully.

TABLE 1.2
Comparative Analysis of Change Management Models for EdTech

Model	Level of Focus	Core Constructs	Strengths in AI-IoT Adoption	Limitations in Educational Context
Lewin's Three-Stage Model	Organizational	Unfreeze-change-refreeze	Clear structural roadmap for staged implementation	Assumes linear change; limited flexibility
Kotter's Eight-Step Model	Organizational & leadership	Vision, urgency, coalition, cultural anchoring	Emphasizes leadership and stakeholder engagement	Resource-intensive; requires sustained commitment
Technology Acceptance Model (TAM)	Individual	Perceived usefulness, ease of use	Explains teacher acceptance behaviour	Limited institutional and policy perspective
UTAUT	Individual & social	Performance expectancy, social influence, facilitating conditions	Accounts for social and infrastructural factors	Less emphasis on long-term cultural embedding
Diffusion of Innovation	Social system	Adoption categories, innovation attributes	Explains spread across institutional networks	Does not provide structured governance strategy

1.7.4 UTAUT and Digital Readiness Models

The Unified Theory of Acceptance and Use of Technology, or UTAUT, is an extension of TAM that accounts for two factors, social influence and facilitating conditions. In educational institutions, such factors as peer endorsement, administrative support and infrastructural reliability are powerful influences on patterns of adoption. Digital readiness models complement UTAUT by examining the preparedness of the organization in terms of infrastructure, governance and human capital. Together these frameworks present a more global attitude towards the systemic adoption capability (Ifenthaler & Yau, 2020).

1.7.5 Diffusion of Innovation in Educational Context

Diffusion of Innovation theory describes the spread of new technologies in social systems through categories of adopters that range from innovators to laggards. In educational spaces, early adopters can play a vital role of change agents within an institutional/social structure to influence the broader acceptance. The rate of diffusion depends on the perceived relative advantage, compatibility and visibility of benefit. This view points out the social dynamics underlying the adoption of AI-IoT systems and raises the importance of peer-led transformations (James, 2025).

Given this diversity in theoretical perspectives, a comparative synthesis of major change management and technology adoption models that are relevant to AI-IoT for education is provided in Table 1.2.

The systemic interaction of AI, IoT infrastructural, data analytics and smart campus operations in an integrated educational ecosystem is observed in Figure 1.2, which shows the collective view of interconnected technologies for adaptive and data-driven institutional transformation.

1.8 STAKEHOLDER PERSPECTIVES AND CONFLICT DYNAMICS

The incorporation of AI and the IoT technologies into educational systems is of multi-stakeholder nature. Educational institutions are like (interrelated) ecosystems in the way that teachers, students, parents, administrators and policymakers interact in frameworks of shared governance. The impact of this is the emergence of resistance to technological transformation is not uniform; rather, it is

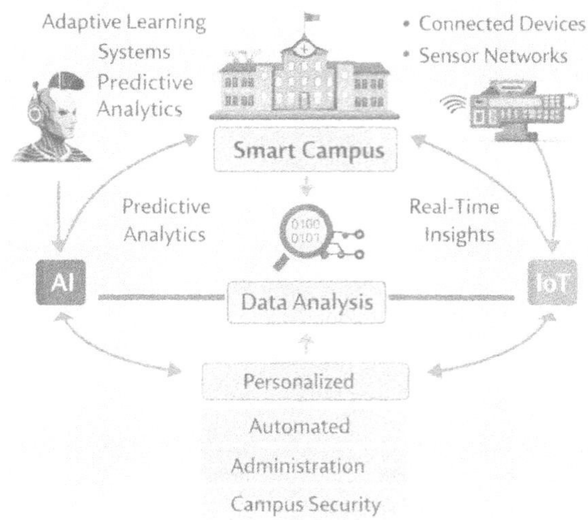

FIGURE 1.2 Integrated AI-IoT adoption model in educational ecosystems.

various expectations, perceived risks and competing priorities. Understanding the stakeholder perspective is an important step for inclusive and sustainable strategies for change. Opposition may frequently stem less from rejection of innovation per such than uncertainty correlated with its implications of autonomy and privacy, accountability and equity. This section keeps the stakeholder-specific concerns in mind and discusses the dynamics of conflicts that govern the trajectory of AI-IoT adoption (Jarrahi, 2018).

1.8.1 Teachers and Academic Staff

Teachers are the main actors from the standpoint of implementing AI-driven and IoT-enabled educational systems. Their acceptance or resistance has considerable bearings on the outcomes of adoption. A huge concern among academic staff is professional autonomy. AI-enabled adaptive learning platforms and automated assessment systems can be viewed as leading to less instructional discretion. When sequencing of content, evaluation standards are affected by algorithmic recommendations, educators may wonder if they are marginalized by the use of their pedagogical expertise.

Another very important issue is the redistribution of workload. While the efficiency gains promised by AI tools are a great opportunity, the early stages of implementation can take more time for AI training, curriculum re-engineering and system troubleshooting. If institutions don't offer the support in an organized manner, there's increased resistance. Academic staff might also indicate issues of data accountability. Performance analytics to track teaching effectiveness has the potential to create anxiety about surveillance and performance benchmarking. However, once the teachers are actively involved in the co-design processes, and given continuous professional development (CPD), resistance is often transformed into constructive engagement. Empowerment and Transparency are key to continuity of teacher trust (Johnson et al., 2020).

1.8.2 Students and Digital Natives

Students, specifically those who are part of the "digital native" generations, are generally more open to the idea of integrating technology. However, receptivity is not the acceptance of unconditionally. To this the students often expressed concerns about data privacy and surveillance. IoT-enabled attendance systems, biometric check and behavioural tracking analytics could be potentially perceived as constant monitoring. The ethical implications of predictive analytics – especially when the goal is to use predictive modelling for performance forecasting or for risk profiling – can raise apprehension. Another concern is that of algorithmic fairness. Students may raise questions about whether AI-based assessment systems accurately reflect the qualitative nature of learning or whether there may be biases embedded in algorithms that could affect academic assessment.

Additionally, the growth of the use of digital learning environments may also lead to an increase in screen dependency and the decrease of face-to-face interactions which may have social developmental implications as well as collaborative learning. Getting students involved in discussions, addressing data governance policies and ensuring transparency with algorithms are thus vital in addressing the opposition (Khanfar et al., 2025).

1.8.3 Parents and Community Concerns

Parents and Community Players are of great importance in shaping public views regarding technological transformation in education. Their fears are often safety, ethics and long-term developmental affects. One issue has been especially prominent – child data protection. Parents, for example, may be concerned about the storage, sharing or commercialization of student data gathered through AI and IoT systems. Without clear consent processes and clear privacy policies, trust deficits can arise. Communities also may get to the question of whether the utilization of AI-driven instruction reduces the interaction with humans, or therefore overly automates, while simultaneously researching humanistic holistic development. Cultural values that focus on teacher-student relationships can make such concerns more pronounced. In addition, socio-economic differences affect parental attitudes. Families from under-resourced backgrounds may be afraid of digital transformation's potential to dumb down gaping inequalities that can exist if access to devices and connectivity are not equally. Proactive communication, public consultations and inclusive digital access initiatives are therefore critical in dealing with resistance at the community level.

1.8.4 Institutional Leadership and Administrators

Institutional leaders are responsible for the strategic planning, allocation of resources and regulatory compliance. Their support is the key to AI-IoT adoption but they also have great operational constraints. One of the biggest concerns of administrators is financial sustainability. AI systems, IoT infrastructure, cyber security frameworks and continuous maintenance, require massive investment. Budgetary constraints can generate some hesitation, especially in public institutions. Leaders also have to deal with reputational and compliance risks. Data breaches, discriminatory algorithms to racial discrimination or systems failures could hurt institutional credibility. Therefore, risk mitigation and legal compliance are a major consideration in decision-making processes. Additionally, there is competition for priorities across departments which administrators must balance. Resistance may emerge if faculty believe that spending on digital investments coming at the expense of traditional academic efforts of the institution in the past. Strategic vision, phased-in and evidence-based evaluation mechanisms can ensure effective strides by institutional leaders in working effectively through these tensions.

1.8.5 Policymakers and Regulators

Policymakers and regulatory bodies exist at the macro-level of the layers of governance. Their role has to do with setting data protection standards, ethical standards for AI use, digital infrastructure development and equal access. Regulators often have the difficult task of finding the balance between promoting innovation and containing risks. Overly restrictive regulations can hamper experimentation, while a lack of oversight can risk exposing students to privacy and/or biased risks. The policy uncertainty may lead to institutional hesitation. Rapidly evolving AI governance frameworks demand institutions to be constantly changing the compliance strategies. Furthermore,

TABLE 1.3
Stakeholder Resistance Matrix and Risk Levels

Stakeholder Group	Primary Concerns	Type of Resistance	Risk Level to Adoption	Mitigation Strategy
Teachers & academic staff	Autonomy loss, workload increase, performance monitoring	Psychological & professional	High	Professional development, participatory design, transparent evaluation
Students	Data privacy, algorithmic bias, surveillance	Ethical & behavioural	Medium-high	Clear data policies, explainable AI systems, feedback channels
Parents & community	Child safety, digital inequality, reduced human interaction	Socio-cultural & ethical	High	Public communication, consent frameworks, digital equity programmes
Institutional leaders	Financial cost, compliance risk, reputational impact	Strategic & operational	High	Phased investment, governance models, risk assessment mechanisms
Policymakers & regulators	Legal accountability, innovation-risk balance	Regulatory & policy	Medium-high	Adaptive regulation, national AI guidelines, stakeholder consultation

national and regional differences in the way regulations are set hamper the crossing national and/or regional borders for academic cooperation. Effective policymaking must therefore incorporate mechanisms for ethical safeguards, Feasibility of Technology, consultation mechanisms with stakeholders, etc. (Pangrazio & Selwyn, 2019).

To get a better understanding of issues raised by stakeholders along with associated implementation risks, a condensed stakeholder resistance matrix is summarized in Table 1.3.

The structural determinations that go into the trust gap between students and AI systems – specifically the perceptions of competence, transparency and privacy can be conceptually demonstrated in Figure 1.3 with respect to perceived algorithmic opacity and data governance issues.

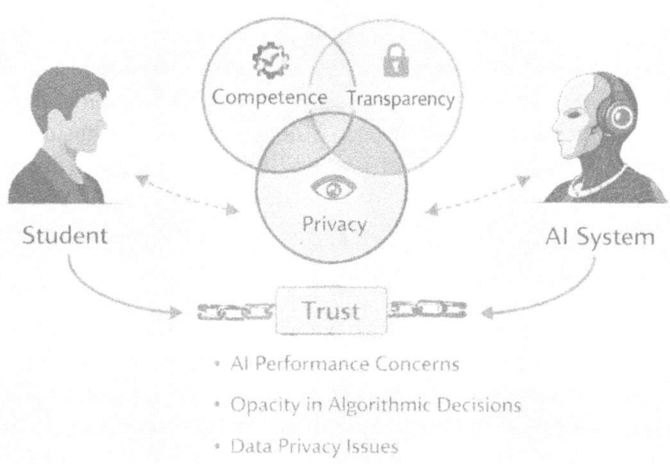

FIGURE 1.3 Trust gap model in AI-enabled classrooms.

1.9 ETHICAL, LEGAL AND GOVERNANCE CHALLENGES

The convergence of AI and IoT technologies in education brings a set of complex ethical, legal and governance considerations that go well beyond the efficiency that these technologies bring when it comes to running process and services. Educational institutions control the sensitive data of our students, and have an impact on long-term development opportunities, so that responsible digital transformation is a matter of public accountability. Resistance to AI-IoT adoption commonly grows when risks that involve privacy, fairness and unclear regulations or social inequalities are perceived by the stakeholders. Therefore, ethical governance ought to be at the heart of technological integration strategies.

1.9.1 Data Privacy and Student Surveillance

AI-driven analytics and IoT-enabled systems are based on unceasing data collection, such as academic performance data, behavioural data, bio identification and real-time engagement data. While there are lessons for using such data that are in favour of personalized learning and institutional planning, there are worries about surveillance and abuse of such data. Students and parents can find constant monitoring to be at best intrusive, especially if they are using facial recognition or location-tracking technology. Effective governance includes explicit consent mechanisms, data minimization policies, encryption standards and transparent retention protocols. Privacy-by-design principles should be the guiding principles for developing the systems to ensure that the student rights are protected from the start (Sairete et al., 2021).

1.9.2 Algorithmic Bias and Fairness

AIs that are trained on history data sets can reproduce structural bias found within datasets. In educational settings, grading, admissions and risk-prediction tests may lead to surprisingly unjust results for minorities through the use of predictive models. To be fair, constant bias checks are needed and training can be in various formats, plus human oversight mechanisms where educators can help examine and challenge the automated decisions. Without such staples, the opacity of algorithms can lead to a lack of legitimacy of the institutions and an increase in resistance.

1.9.3 Transparency and Explainability in AI Systems

The opacity of complicated machine learning models creates difficulties of accountability. When AI systems lead to grading decisions or intervention strategies with no clear explanation, we will see them lose faith in their reliability. Explainable components of AI strive to give interpretable reasoning of the output made by algorithms, which can help educators and administrators gain an understanding of, and contest, a decision where to the matching a decision might be necessary. Transparent documentation and reporting mechanisms and avenues of appeal can enhance trust and even scepticism towards automated systems (Shankar et al., 2025).

1.9.4 Global Regulatory Developments (2020–2035 Outlook)

Since 2020, there has been a greater focus in regulators worldwide around protecting data, taking algorithms responsible and implementing ethical AI. Regulatory frameworks based on risks, mandatory impact assessments and heightened data protection laws are changing institutional compliance requirements. Looking down the road to 2035, harmonization of international governance standards in the sphere of AI is likely to have an impact on educational collaboration conducted between countries and planning of digital infrastructure across borders. Therefore, it is critical for institutions to implement adaptive compliance strategies that can accommodate to changing regulation expectations and not stop the pace of innovation (Stahl, 2025).

TABLE 1.4
Global AI-in-Education Regulatory Landscape

Region	Key Regulatory Focus	Data Protection Strength	AI Governance Mechanisms	Educational Impact
European Union	Risk-based AI regulation, GDPR compliance	Very high	Mandatory AI risk assessment, transparency obligations	Strong privacy safeguards; compliance-intensive deployment
United States	Sectoral data laws, state-level AI guidelines	High (varies by state)	Algorithmic accountability initiatives, FERPA protections	Innovation-driven but regulatory fragmentation
India	Digital personal data protection framework, national AI strategy	Moderate-high	Emerging AI governance policies, digital infrastructure expansion	Rapid digital adoption with evolving safeguards
China	State-led AI governance and data localization	High (state-regulated)	Centralized AI oversight, smart campus initiatives	Large-scale implementation with strong regulatory control
Middle East (UAE, Saudi Arabia)	National AI strategies, smart education initiatives	Moderate-high	Government-led AI adoption frameworks	Accelerated innovation with centralized oversight

1.9.5 Digital Equity and Inclusion

Digital transformation risks will increase socio-economic inequalities if equitable access to infrastructure and digital literacy is not achieved. Unequal access to devices, connectivity and limited strategies of accessibility contribute to education inequality. Inclusive AI design, subsidized connectivity programmes and universal access policies are hence critical constituents of responsible adoption. Embedding equity considerations into the strategies for procurement and implementation means that the advancement of technology will create much needed enhancement of educational documentation rather than its detrimental undermining. ancillary as it may be, the ethical, legal and governance challenges indicate that the integration of AI and IoT needs to be grounded in princely rules of legal and regulatory oversight, transparency, accountability and inclusive policy making to ensure a sustainable and socially legitimate transformation.

In view of the changing regulatory scenario in different parts of the world, a comparative picture of AI regulatory frameworks in various major regions is given in Table 1.4.

A structured framework describing key areas of ethical risk (bias and fairness, accountability, privacy and security and sustainability) and corresponding mitigation strategies for AI and IoT deployment in schools are illustrated in Figure 1.4 with a focus on a comprehensive and preventive governance approach.

1.10 STRATEGIC FRAMEWORKS TO OVERCOME RESISTANCE

Overcoming resistance to AI and IoT integration in education needs a structured and multidimensional strategic approach. Technological transformation cannot only happen by infrastructure investment or ordered through policy decisions, but must have supportive institutional readiness, stakeholder empowerment, governance clarity and sustainable professional development. Resistance is very often uncertainty, perceived risk or misalignment of institutional culture and digital ambition. Therefore, strategic frameworks have to consider structural and psychological aspects of change to guarantee sustainable adoption.

FIGURE 1.4 Ethical risk mitigation framework for AI and IoT in schools.

1.10.1 Institutional Change Readiness Assessment

Effective transformation starts with taking a critical assessment of institutional readiness. Educational institutions vary greatly in terms of their technological maturity, leadership orientation and digital maturity. A systematic assessment of readiness looks at the digital infrastructure robustness, availability of cybersecurity safeguards, level of digital literacy among the faculties and visibility of alignment of institutional policies with AI governance requirements. Without such preliminary evaluation, implementation of technology can create new vulnerabilities for the organization and increase resistance.

Institutional culture is also the deciding factor. If leadership cannot set out a coherent vision of digital intelligence (AI-IoT), AI-IoT efforts may be seen as being fragmented or imposed from the outside. Readiness assessment thus is not only playing the technical audit role but also serves as a strategic diagnostic tool to understand the gaps in capabilities and for phased implementation (Tang et al., 2022).

1.10.2 Capacity Building and Continuous Professional Development

Sustainable AI-IoT integration has at its core the capacity of human beings. Faculty members need to be familiar not only with how to do something in digital systems, but also with the pedagogical implications and ethical dimensions of those systems and their usage. CPD initiatives should therefore go beyond technical orientation and include AI literacy, data interpretation skills among other support with digital pedagogy training and cybersecurity awareness. Professional development must be organized as a continuous process as opposed to a one-time intervention. As the technology of AI is evolving at an abundant rate, educators need the adaptability learning opportunities where we can keep ourselves competent and confident. Investing in mentoring systems, collaborative innovation workshops and interdisciplinary knowledge exchange create a culture of technological ownership in institutions investing in these areas. When educators feel that they are active participants in the, instead of passive recipients of, change, psychological resistance is greatly reduced (Teräs, 2022).

1.10.3 Participatory and Co-Creation Approaches

Resistance increases when technological decisions are introduced through top-down hierarchical processes. Participatory governance models are a more inclusive way of transformation. Involving teachers, students, parents and technical staff in consultation processes helps to strengthen institutional legitimacy and increases trust. Co-creation approaches hold that stakeholders can offer input into the selection of technologies and implementation planning as well as in the ethical oversight of these facets. Faculty committees can verify pedagogical fit; student advisory committees can give feedback on user and privacy issues. Such participatory structures ensure not only better quality in decision-making, but also create collective responsibility. By turning stakeholders into partners in innovation, the institutions actually turn opposition into cooperative involvement.

1.10.4 Transparent Communication Strategies

Transparency of communication is the key to overcoming uncertainty-based resistance. For AI-IoT deployment, educational institutions need to clearly define the goals and expected results and risk mitigation strategies. Ambiguity surrounding the use of data, algorithmic decision-making or monitoring practice can become a source of distrust and resistance. Comprehensive communication strategies in the form of regular progress updates and availability of data governance policy have been made publicly as well as open forums to talk to stakeholders. Transparency should also apply to algorithmic processes so that decisions that are influenced by AI systems are interpretable and contestable. When stakeholders are knowledgeable about how technologies function and the extent to which accountability is ensured, institutional trust is enhanced and resistance is reduced (Williamson & Eynon, 2020).

1.10.5 Governance and Accountability Mechanisms

Enabling Ethical and Lawful AI Infrastructure Effective governance structures are required to provide ethical and legal integration of AI-IoT. Institutions are required to implement clear structures of accountability, through the provision of designated data protection officers, and AI ethics committees and independent mechanisms of review. Periodic Audits of the algorithm are performed as well as cybersecurity risk assessments to reinforce institutional responsibility and reduce vulnerability in operational efficiencies. Governance mechanisms should also define distribution of responsibilities among the institutional leadership, IT departments and third-party vendors. Transparent contractual agreements about data ownership, use rights and about liability terms are important to protecting institutional autonomy. By including accountability in the structure of governance, institutions signal their sense of responsibility towards ethical subversion of digital transformation processes and help alleviate scepticism among stakeholders (Jarrahi, 2018).

1.10.6 Public-Private Collaboration Models

Given the complexity and cost of AI-IoT systems, it may be unavoidable to work with technology providers and industry partners. Public-private partnerships can help to make advanced infrastructure, technical know-how and research innovation available. However, there needs to be regulations on the implementation of such collaborations that are based on transparent procurement policies and strict data protection to avoid any commercial exploitation of educational data. Strategically structured partnerships might involve shared research efforts, innovation testing environments and common cybersecurity structures. When tied to public interest goals, much faster technology development can be achieved with institutional accountability. Effective partnership governance therefore changes external dependency to strategy synergy.

To support phased implementation as well as strategic, long-term planning, a roadmap for AI and IoT integration is presented in Table 1.5 for 2025 through 2035.

A structured four-phase progression of exploring and planning – pilot deployment – scaling and optimization – and sustainability evaluation for systematic integration of AI-IoT in educational institutions is demonstrated in Figure 1.5 – developing systematic capacity building and governance alignment across the lifecycle of implementation.

TABLE 1.5
Strategic Roadmap for AI and IoT Integration (2025–2035)

Phase	Timeframe	Strategic Focus	Key Institutional Actions	Expected Outcome
Phase I	2025–2027	Readiness & pilot implementation	Infrastructure audit, faculty training, small-scale pilots	Reduced uncertainty and initial trust building
Phase II	2027–2030	Scaled deployment & governance strengthening	Policy refinement, AI ethics committees, expanded IoT integration	Institutional alignment and operational stability
Phase III	2030–2035	Optimization & sustainable integration	Continuous monitoring, advanced analytics, equity-focused expansion	Fully integrated, resilient digital ecosystem

1.11 GLOBAL CASE STUDIES AND Best PRACTICES

The overall scenario of AI and IoT integration in education is characterized by multiple pathways of implementation, dependent on the cultural, economic and regulatory context. While the countries have taken different paths, some have developed comprehensive national AI education strategies, while others have invested infrastructure in modernization or have produced results through pilot education. Looking at experiences in different countries of the world will be a source of useful insights into how to overcome resistance, build more effective governance and ensure sustainable change. This section introduces some of the regional case illustrations and outlines some of the lessons that can be transferred to other institutional contexts.

1.11.1 AI Curriculum Integration in Nordic Countries

Nordic countries have become the pioneers in integrating AI literacy into the national curriculum. For instance, projects at Finland have placed emphasis on mathematically oriented thinking,

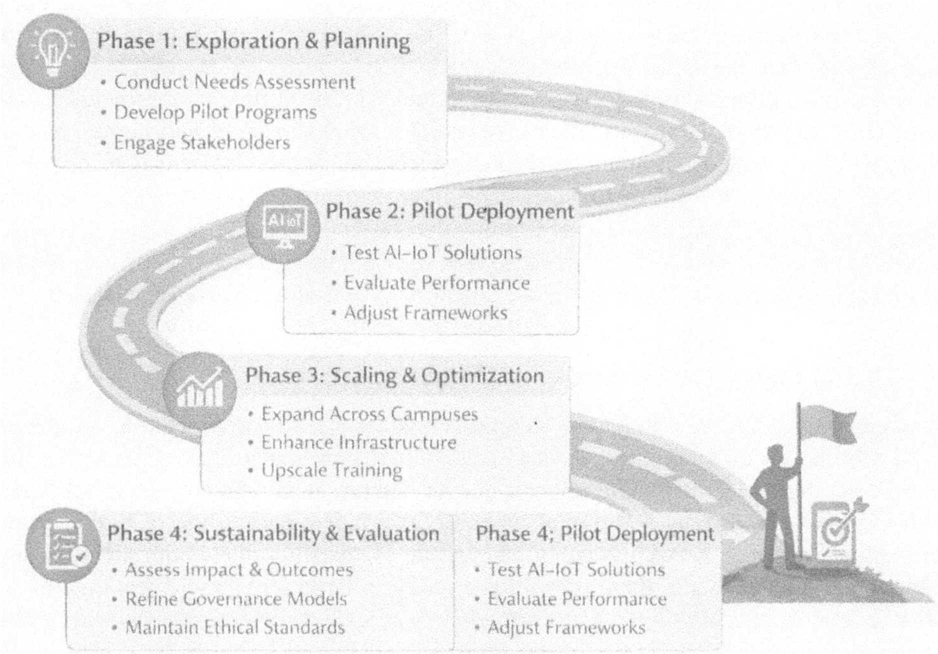

FIGURE 1.5 Phased implementation roadmap for AI-IoT adoption.

awareness of ethics and interdisciplinary digital competence, from elementary school onwards. Rather than centring AI as a technical topic, curricular reforms combine algorithmic literacy in the areas of humanities, social sciences and STEM disciplines. This type of holistic approach has decreased the reaction of resistance because it is able to frame AI education as a societal competency instead of a disruptive technological intrusion.

Teachers are given structured programmes of professional development in line with national digital strategies. In addition, high public confidence in the governance institutions helps the public accept data-driven educational reforms. A major success feature lies in implementing incrementally. As opposed to sharp technological breaks, Nordic systems focus on curriculum restructuring, teacher empowerment and ethical discourse, so as to incorporate AI into pedagogical philosophy rather than as an external technological tool (Yu et al., 2024).

1.11.2 Smart Campus Implementation in East Asia

The education systems in East Asia have spent heavily on smart campus ecosystems, which are IoT-enabled. Countries like China have invested in large-scale smart classroom network projects that include facial recognition attendance system, environmental sensors and AI-based learning analytics and centralized data platform for the campus. These are implementations that show how centralized governance structures can speed up technological implementation. However, they also suggest the importance of regulatory oversight and cybersecurity safeguards. In technologically smart cities, IoT integration has enabled greater efficiencies in resource use, campus security and administrative coordination. The East Asian model provides a good example of the close relationship between infrastructure readiness and policy coherence and their effects on the adoption speed. Nevertheless, the engagement of stakeholders remains essential to reducing issues surrounding surveillance and data control.

1.11.3 National AI Education Policies in the Middle East

Middle Eastern nations have started to make AI one of the key pillars of national growth. For some, national AI strategies have been launched in the United Arab Emirates around building digital skills, promoting smart education platforms and innovation in public industries.

Educational institutions within the region owe to the centralized funding mechanisms and government orchestrated digital transformation agendas. This top-down model helps in fast modernization of infrastructure such as AI-augmented adaptive learning systems, IoT enabled smart learning environments. However, for successful sustainable development, there also needs to be alignment between technology spending and teacher training debates and ethical governance. It is important that regions embracing tough policy towards AI find the right balance between driving innovativeness and institutional capacity building for the long term, avoiding surface-level adoption.

1.11.4 Digital Infrastructure Reforms in Emerging Economies

Emerging economies also tend to have structural issues to contend with, including limited access to broadband, uneven distribution of devices and limited financial resources. Countries such as India have done digital infrastructure reforms that go hand in hand with larger national education policy to increase connectivity and digital literacy. Government-backed use of digital platforms, open educational resources as well as public and private partnerships play important roles in bridging infrastructural gaps. While there is still uncertainty in large-scale implementation of AI, the gradual efforts to implement digital transformation show us how emerging economies can adopt phased adoption models. Important to these reforms is the emphasis on equity centric strategies. Subsidized programmes on connectivity and the availability of devices solve the issue of digital divide by changing the socio-economic resistance and facilitates the inclusive adoption.

1.11.5 Lessons Learned from Early Adoption Failures

All AI-IoT integration efforts have not necessarily been successful over the long term. Early adoption failures in different parts of the world offer lessons to be learned from. Common pitfalls: Failure to consult other stakeholders, Teachers are not adequately trained to use the vendors, Resources: High dependence on discrimination, Gender: Cybersecurity is lacked. Backlash due to privacy or bias concerns common for institutions which implemented AI-based systems without clear governance systems. Similarly, IoT deployments without a maintenance plan resulted in operational inefficiencies and discontinuation. These experiences point to a number of principles of great importance:

First, technological implementation should be founded on pedagogical goals rather than administrative efficiency goals.
Second, stakeholder trust should be built by transparency and participatory governance.
Third, infrastructure investment needs to be accompanied by professional development that is sustained over time.
Fourth, regulatory compliance and ethical oversight is necessary prior to large scale deployment (Yoo et al., 2020).

Collectively, the case studies from throughout the world show that the resistance can be tamed where transformation is strategic, inclusive and has an ethical base. Successful integration of AI-IoT is not more dependent on technology blockchain but rather depends more on governance coherence, trust building for stakeholders and phased alignment of institutions (Yu et al., 2024).

1.12 FUTURE DIRECTIONS (2025–2035)

The metamorphosis of AI technologies and IoT technologies from 2025 to 2035 is projected to alter the educational transformation course beyond the early stage of adoption to deeply embedded intelligent ecosystems. As resistance evolves into strategic adaptation, an increase from implementation challenges to optimization, refinement of ethics and sustainability will be the focal point. The next decade would mostly see the rise of collaborative intelligence models, decentralized AI architectures, predictive governance systems and green digital infrastructures. Anticipating these trajectories is key to institutions trying to future proof their transformation strategies.

1.12.1 Human-AI Collaborative Teaching Models

The future AI in education will not be about teacher replacement, but rather teacher augmentation. Human-AI collaborative teaching models are expected to be a combined model of adaptive learning algorithms and human pedagogical judgment, which will serve to improve the process of instructional personalization, while maintaining sleep relationships of instruction. In such models, AI systems analyse patterns and make recommendations about instruction and identify at-risk students, while educators interpret the nuances of a situation, emotionally support students and use ethical discretion. This collaborative paradigm will revise professional roles. Teachers may have to increasingly be identified as wards of learning architects and data-informed mentors beyond the role of being the sole content transmitter. Resistance may subside as the power of AI tools will make them realize their ability to lessen administrative burden and deliver differentiated instruction. However, Venice will have to collaborate successfully and some AI literacy training and clear delineation of the authority to make decisions will have to occur so that we do not over-rely on automated recommendations. Giving centre-stage to the human agency a fundamental element of pedagogical practice will be vital to sustain institutional trust (Zhang & Zhang, 2024).

1.12.2 Edge AI and IoT-Enabled Smart Campuses

The increased use of IoT devices, at educational campuses, will lead to an increased trend in edge computing architectures. Edge AI distributes processing power and operations to the edge of the system either in devices or in order to minimize the amount of time required to run the system. Between 2025 and 2035 smart campuses are likely to include integrated sensors to monitor the environment, adaptive energy management systems, real-time monitoring of campus security systems and smart resource allocation platforms.

Edge AI integration will be able to speed up decision-making in case of classroom optimization, emergency response and infrastructure maintenance. Importantly, decentralized data processing alternatives can also address privacy concerns to limit the ability to aggregate large amounts of data. However, the coming expansion of IoT ecosystems will need strong cybersecurity protocols and interoperability frameworks. The robustness of institutions during the rise or growing technological complexity will be at an advantage as those investing in resilient network architectures and encoding mechanisms are better able to handle quantitative or quantitative complexity.

1.12.3 Predictive Analytics for Institutional Decision-Making

Predictive analytics will likely become a major part of institutional governance. By using machine learning algorithms, educational industries can anticipate enrolment trends, detect dropout probabilities and manage efficient resource allotment and foresee alterations in demand for curriculum offerings. Such capabilities make operation more strategic and more efficient.

However, a dependence on predictive systems does raise the issue of governance in addition to a question of ethics. Forecasting models need to be continually audited so as to guard against bias amplification and fairness in student intervention strategies. Institutions need to balance to make a decision between data analytics and other factors such as judgment and ethics. Transparent documentation of predictive methodologies will continue to be vital to the maintenance of stakeholder confidence. The maturation of predictive analytics could also be supportive of evidence-based policymaking at the country and regional supremacy, aiding governments to plan funding and infrastructure investments more correctly. Nonetheless, predictive governance needs to be accountable to human processes of review in order to avoid overreach by technology.

1.12.4 Sustainable and Green AI Deployment

As the use of AI and IoT continues to expand around the world, there will be an increased focus on environmental sustainability issues. AI systems, especially those that are used for data processing on a large scale and with the help of cloud computing, require a lot of energy resources. Between 2025 and 2035, the deployment of sustainable and green AI will be one of the strategic priorities in educational space. Energy efficient edge computing, renewable power-based data centres and algorithm optimization will play an important role in lowering carbon footprints. Smart campus infrastructures may have IoT-based energy monitoring systems that adjust the usage of resources dynamically. Integrating sustainability metrics into procurement policies and into the process of managing vendors will lead to environmentally responsible digital transformation. Moreover, co-ordination of sustainability education in AI-enabled curricula can link the technological development to the environment. Institutions that pursue green AI strategies not only save cost in their operations but they can also increase social legitimacy through their actions of dedicating to global sustainability goals.

1.13 CONCLUSION

The intertwining of AI and IoT technologies in education is a game-changer evolution reaching beyond technical innovation and institutional and ethical and cultural transformations. As discussed throughout this chapter, however, resistance to such transformation is not only natural but multidimensional, suffice it to conclude that resistance to such transformation arises due to concerns related to professional autonomy, data privacy, governance accountability, infrastructural readiness and socio-economic equity. Addressing these concerns requires something more than technological implementation, it requires the management of change in a structured way based on strategic planning and engagement with stakeholders.

By bringing together synthetic concepts of theoretical models, global case studies and the principles of governance, this chapter proposed the Unified Change Management Framework as a sustainability framework that would support AI-IoT integration. The key elements at the heart of this framework are institutional readiness assessment, participatory implementation, the open flow of communication, solid governance and the ongoing implementation of monitoring. These elements all together help to reduce uncertainty and build trust among the stakeholders. Looking forward to the time period of 2035, the coming of human-AI collaboration, predictive analysis, smart campus ecosystems and sustainable digital practices were to define the next phase of educational transformation. Institutions that link innovation and ethical responsibility as well as inclusive policies and design will be best equipped to transform resistance into resilience in order to guarantee long-term digital sustainability.

REFERENCES

Akgun, S., & Greenhow, C. (2022). Artificial intelligence in education: Addressing ethical challenges in K-12 settings. *AI and Ethics*, *2*(3), 431–440.

Alhumaid, K., Ali, S., Waheed, A., Zahid, E., & Habes, M. (2020). COVID-19 & e-learning: Perceptions & attitudes of teachers towards e-learning acceptance in the developing countries. *Multicultural Education*, *6*(2), 100–115.

Bender, E. M., Gebru, T., McMillan-Major, A., & Shmitchell, S. (2021). On the dangers of stochastic parrots: Can language models be too big? In *Proceedings of the 2021 ACM conference on fairness, accountability, and transparency* (pp. 610–623).

Bond, M., Bedenlier, S., Marín, V. I., & Händel, M. (2021). Emergency remote teaching in higher education: Mapping the first global online semester. *International Journal of Educational Technology in Higher Education*, *18*(1), 50.

Bond, M., Khosravi, H., De Laat, M., Bergdahl, N., Negrea, V., Oxley, E., … Siemens, G. (2024). A meta systematic review of artificial intelligence in higher education: A call for increased ethics, collaboration, and rigor. *International Journal of Educational Technology in Higher Education*, *21*, 1.

Cabrera, B. M., Luiz, L. E., & Teixeira, J. P. (2025). The Artificial Intelligence Act: Insights regarding its application and implications. *Procedia Computer Science*, *256*, 230–237.

Cannarsa, M. (2021). Ethics guidelines for trustworthy AI. *The Cambridge Handbook of Lawyering in the Digital Age*, *30*, 97–283.

Chan, C. K. Y., & Colloton, T. (2024). *Generative AI in higher education: The ChatGPT effect.* London: Routledge.

Chassignol, M., Khoroshavin, A., Klimova, A., & Bilyatdinova, A. (2018). Artificial Intelligence trends in education: A narrative overview. *Procedia Computer Science*, *136*, 16–24.

Dwivedi, Y. K., Hughes, L., Ismagilova, E., Aarts, G., Coombs, C., Crick, T., … Williams, M. D. (2021). Artificial Intelligence (AI): Multidisciplinary perspectives on emerging challenges, opportunities, and agenda for research, practice and policy. *International Journal of Information Management*, *57*, 101994.

Floridi, L., & Cowls, J. (2022). A unified framework of five principles for AI in society. *Machine learning and the city: Applications in architecture and urban design* (pp. 535–545).

Gunasekara, L., El-Haber, N., Nagpal, S., Moraliyage, H., Issadeen, Z., Manic, M., & De Silva, D. (2025). A systematic review of responsible artificial intelligence principles and practice. *Applied System Innovation*, *8*(4), 97.

Holmes, W., & Porayska-Pomsta, K. (2023). *The ethics of artificial intelligence in education.* London: Routledge, 621–653.

Ifenthaler, D., & Yau, J. Y. K. (2020). Utilising learning analytics to support study success in higher education: A systematic review. *Educational Technology Research and Development*, *68*(4), 1961–1990.

James, D. (2025). The impact of artificial intelligence on teaching and learning in higher education. *Actual economy: Afro-Asian solutions to global challenges-2025* (pp. 354–361).

Jarrahi, M. H. (2018). Artificial intelligence and the future of work: Human-AI symbiosis in organizational decision making. *Business Horizons*, *61*(4), 577–586.

Johnson, N., Veletsianos, G., & Seaman, J. (2020). US Faculty and Administrators' experiences and approaches in the early weeks of the COVID-19 pandemic. *Online Learning*, *24*(2), 6–21.

Khanfar, A. A., Kiani Mavi, R., Iranmanesh, M., & Gengatharen, D. (2025). Factors influencing the adoption of artificial intelligence systems: A systematic literature review. *Management Decision*, *63*(10), 3727–3755.

Pangrazio, L., & Selwyn, N. (2019). 'Personal data literacies': A critical literacies approach to enhancing understandings of personal digital data. *New media & Society*, *21*(2), 419–437.

Sairete, A., Balfagih, Z., Brahimi, T., Mousa, M. E. A., Lytras, M., & Visvizi, A. (2021). Artificial intelligence: Towards digital transformation of life, work, and education. *Procedia Computer Science*, *194*, 1–8.

Shankar, R. S., Shailesh, T., R, V. A., Chadaga, K., & Cenitta, D. (2025). Smart health monitoring: An interactive Raspberry Pi-based system for managing diabetes and hypertension in children. *Journal of Cloud Computing*, *14*(1), 70.

Stahl, B. C. (2025). The ethics of data and its governance: A discourse theoretical approach. *Information*, *16*(6), 497.

Tang, Q., Xie, R., Yu, F. R., Chen, T., Zhang, R., Huang, T., & Liu, Y. (2022). Distributed task scheduling in serverless edge computing networks for the internet of things: A learning approach. *IEEE Internet of Things Journal*, *9*(20), 19634–19648.

Teräs, M. (2022). *Education and technology: Key issues and debates: Neil Selwyn*. London and New York: Bloomsbury Academic. 2022, 222 pp. ISBN 978-1-3501-4554-2 (hbk), ISBN 978-1-3501-4555-9 (pbk), ISBN 978-1-3501-4553-5 (ePDF), ISBN 978-1-3501-4556-6 (eBook).

Williamson, B., & Eynon, R. (2020). Historical threads in AI in education. *Learning, Media and Technology*, *45*(3), 223–235.

Yoo, I., Kim, W., Jeon, J., Yu, W., & Bae, Y. (2020). A study on the future directions according to analysis of necessity of AI education. *Journal of The Korean Association of Information Education*, *24*(5), 423–431.

Yu, J., Albashrawi, M., & Tarhini, A. (2024). The determinants of the loyalty of mobile banking: The integrated model of IS success model and UTAUT. *Pacific Asia Journal of the Association for Information Systems*, *16*(2), 3.

Zhang, J., & Zhang, Z. (2024). AI in teacher education: Unlocking new dimensions in teaching support, inclusive learning, and digital literacy. *Journal of Computer Assisted Learning*, *40*(4), 1871–1885.

2 Beyond the Hype

Artificial Intelligence in Education

Yahya Fikri, Rocky Kumar, Bhanuprakash Madupati, Abdelhamid Nechad, Mohamed Rhalma, and Redouane Kaiss

2.1 INTRODUCTION

AI has become one of the most powerful technologies that is affecting the current education systems. From personalised learning platforms to intelligent tutoring systems (ITS), automated assessment systems and learning analytics, as AI tools go, there is an increasing level of embedding into different levels of education. Governments, institutions and EdTech companies propagate the "AI" solution widely to solve long-standing problems of education, such as scalability, learner diversity, efficiency, etc. As a result, AI has been described in many ways as a teaching and learning practices transformational force that has the potential to redefine teaching and learning practices. However, with opportunities that provide real technological advances, also there is an over-optimistic painting of our new technology for education, and expectation. Marketing assertions often lead to empirical facts and the nuts and bolts of how the learning material will be adopted in a classroom, but institutional readiness and ethical responsibility go unaddressed. This chapter argues that it is necessary to critically and evidence-based look at AI in education so that we are no longer stuck on a surface level of enthusiasm and begin making a more significant education impact (Aldenaini, 2021).

2.2 RISE OF AI IN EDUCATION

The use of AI in education has been accelerated thanks to the progress in machines learning, accessibility to education data and digitalisation of the learning environment in mass. Online learning platforms, learning management systems and digital assessment tools are generating mammoth quantities of data to analyse in the human realm behaviour and performance of learners through an AI utility (Almaskari et al., 2021). Educational institutions are utilising AI fast to make learning more adaptive, automated grading, predictive learning by AI. These developments have led to the fact that AI has become an integral part of the EdTech Ecosystem of today. While such technologies promise promising capabilities, the effectiveness of such technologies is dependent on contextual factors such as the pedagogical alignment, the teacher involvement and institutional infrastructure.

2.3 WHY "BEYOND THE HYPE" MATTERS IN EDTECH

The phrase "Beyond the Hype" is putting in perspective the need to look back on AI claims in Education. Many AI-based educational tools are being touted as revolutionary with no proper validation with classroom-based facts, and always with no long-term study. This cases this gap between the promise and proof can lead to the investments of wrong technologies, unrealistic expectations from educators, which can lead to resistance in the adopting of educational technology. A hype-driven approach is attractive as it has the risk of making the tech-fad novel, rather than valuable from the pedagogical perspective. By contrast, an evidence-based perspective focuses on obtainable outcomes of learning, support of teachers and ethical responsibility. Moving beyond the hype is key

DOI: 10.1201/9781003767855-2

to helping stakeholders differentiate between potentially useful AI applications and those that do not provide much educational value (Han et al., 2024).

2.4 GAP BETWEEN AI PROMISES AND CLASSROOM REALITY

Despite the substantial talk about the potential of AI, its process of being applied into the day-to-day practice in the classroom is uneven. Teachers frequently deal with problems concerning the quality of data, transparency of algorithms, lack of training and lack of control over AI-based systems. Further, infrastructural limitations and differences in the context of various educational settings all have an impact on the effectiveness of AI solutions. In the most part, one can see AI tools introduced without due thought of classroom dynamic, learner diversity, or alignment with curriculum. The human-centred approach in design and evidence-based evaluation become important in this context to help AI to be used as a tool for supporting teachers instead of disrupting their work, which is what this gap quite literally points to (Chassignol et al., 2018).

2.5 OBJECTIVES, SCOPE AND STRUCTURE OF THE CHAPTER

The main goal of this chapter is to give a balanced and critical evaluation about AI in Education trying to go beyond exaggerated statements towards specific and ethical-pedagogical insights. The chapter is an attempt to bridge the divide between technological terminology and educational reality.

Among the aspects covered by the chapter are:

- Evolution and fundamental technologies of AI in education
- Practical uses and limitations of educational tools based on Artificial Intelligence
- Ethical, privacy and governance issues
- Real-world examples of cases with evidence of real-world AI adoption

The chapters are designed so they progress in introducing the role of AI in education from the basics to move on to practical and ethical discussions and future-oriented topics.

NOVELTY BOX 1: WHY THIS CHAPTER IS UNIQUE

This chapter is unique in that it features a critical, evidence-based view about artificial intelligence in education. Rather than hyping AI as a then universal solution, however, it assesses real implementations, points to limitations and highlights the importance of educators and pedagogy continuing in the role of being important. Using the integration of technological analysis and governance through ethics and institutions, the chapter offers the framework of an adopted human-centred approach to create a human ethical framework for the adoption of AI in education.

As abridged in Table 2.1, evidence-based AI adoption puts pedagogical effectiveness and ethical responsibility – not hype-driven innovation – as the prioritised aspects of AI adoption, along with teacher involvement.

2.6 EVOLUTION OF AI IN EDUCATION

The use of artificial intelligence in education has undergone a number of decades of change, and has been influenced by improvements in computing power, data availability and pedagogical theory. Rather than a single disruptive technology, AI in education has evolved in a distinctly

TABLE 2.1
Hype-Driven vs Evidence-Based AI Adoption in Education

Aspect	Hype-Driven AI Adoption	Evidence-Based AI Adoption
Primary focus	Marketing claims and innovation appeal	Learning outcomes and pedagogy
Validation	Limited empirical evidence	Classroom-tested and data-supported
Role of teachers	Often marginalised	Central and collaborative
Decision-making	Vendor-driven	Educator- and policy-driven
Ethical considerations	Often overlooked	Explicitly addressed
Sustainability	Short-term enthusiasm	Long-term educational value

different way – through distinct phases – each of which is characterised by different technical capabilities, educational objectives and limitations. Understanding this evolution is crucial in order to understand the context of current AI-driven educational systems, in order to estimate their sensible potential (Chen et al., 2020).

2.6.1 Early Computer-Assisted Learning and Expert Systems

The first application of AI related concepts to schooling can be drawn backbone to processer helped knowledge (CAL) systems developed in the 60s and 70s. These systems were largely rule-based and were designed to deliver instructional content in a linear and predefined way. Even though they are not intelligent in the way we would think today, they provided the basis for mechanised teaching and learning at your own pace.

During the 1980s and early 1990s, expert systems were the first explicit incorporation of artificial intelligence into education environments. These systems were based on handcrafted rules and domain knowledge for providing tutoring and feedback especially in well-structured subjects, like mathematics and programming. ITS emerged out of this paradigm, which sought to model the learner's knowledge and adapt to it.

Despite their great conceptual importance, early expert systems had major limitations. They were costly to develop, not easily scalable and not flexible. Most importantly, they relied very heavily on predefined rules and were therefore incapable of learning from data and adapting to the variety of learner behaviours (Douglass et al., 2024).

2.6.2 Shift Towards Machine Learning and Learning Analytics

The end of the 1990s and the early 2000s saw a slow shift from rule-based AI to machine learning (ML) approaches from education. This transition was prompted by the growing level of digitisation of the educational processes and the prevalent custom of knowledge running systems (LMS). These platforms produced vast amounts of learner data within the platforms, making them capable of data-driven analysis of student behaviour and student performance (Dwivedi et al., 2021).

Learning analytics has become one of the main areas of application that focuses on the usage of ML methods to sense the designs in educational information. Predictive models have been created to forecast schoolchild recital, recognise at-risk students, fold in institutional decision-making. Unlike expert systems, ML-based approaches are capable of learning from past datasets and getting better in time. However, within this phase, there were also some new challenges such as problems with data quality, lack of transparency in predictive models and student privacy concerns. While ML-based learning analytics levels of learning analytics had more scalability than what the previous learning analytics system had to offer, its educational impact was often tied to the interpretation and action on insights from educators.

2.6.3 Emergence of Adaptive and Intelligent Learning Platforms

The development of adaptive and intelligent learning platforms has been made possible cheers to new fees in profound knowledge, natural language processing and modern advancements in computational power. These systems are being used to personalise the learning experience – that is, dynamically modifying the content, pace and feedback based on the individual learner profiles.

Contemporary platforms offer features such as adaptation of learning pathways, automated assessment and conversational interfaces (chatbots and virtual teaching assistants). Unlike previous systems, these platforms can process unstructured data, such as text and interaction logs so that learner behaviour can be richly modelled. While adaptive platforms are a major technological advancement, they are not equally effective in education. Successful implementation depends on the alignment with curriculum objectives, teacher involvement and ethics protection. The development points to a recurring theme that AI in education-making is blind optimisation of technology does not guarantee meaningful learning outcomes.

As summarised in Table 2.2, the journey of AI in education is a shift from rule-based systems in education towards data-driven and adaptive learning platforms that have clearly defined capabilities and limitations.

2.7 CORE AI TECHNOLOGIES USED IN EDUCATION

The effectiveness of artificial intelligence in education depends wildly on the technologies that support it: Data analysis, personalisation and automation. Rather than being a monolithic system, AI in education is manifested through a combination of ML, natural language processing and computer vision (CV) techniques. Each of these technologies is designed to meet particular educational needs, spanning from analytics and assessment for learning, engagement monitoring and behavioural analysis. This section explores the underlying technologies of AI that are and are gaining acceptance in educational systems and analyses their practical role and limitations.

2.7.1 Machine Learning and Deep Learning for Learning Analytics

ML is at the centre of most development of an AI-based educational system, especially in the concept of learning analytics. ML algorithms are used to understand how learning trends, performance and behaviour relate to educational outcomes by analysing vast amounts of educational data such as educational assessments and recordings of learning activities and interactions as well as attendance records.

Traditional ML models such as regression and decision trees are common methods used in order to determine success in school and predict which students are at risk. More recently hot is deep learning (DL) techniques because of their capacity to record complex, no linear data relationships in

TABLE 2.2
Evolution of AI Technologies in Education and Their Capabilities

Era	AI Technology	Primary Educational Application	Key Capabilities	Major Limitations
1960s–1970s	Computer-assisted learning	Programmed instruction	Automated content delivery	No adaptivity
1980s–1990s	Rule-based expert systems	Intelligent tutoring	Domain-specific guidance	Poor scalability
2000s	Machine learning	Learning analytics	Data-driven prediction	Limited interpretability
2010s–present	Deep learning and NLP	Adaptive learning platforms	Personalisation, automation	Ethical and privacy concerns

high-dimensional data. Neural networks are now starting to be applied in the modelling of learning trajectories, content delivery or for systems implementing adaptive learning.

While the ML and DL can increase the predictive accuracy and the scalability of the process, it is only through data quality, interpretability and by making valuable sense of it in pedagogical decision-making that the method becomes effective for education. Without proper contextualisation, data-driven insights about analytics may not be translated into better learning and outcomes (Holstein et al., 2019).

2.7.2 Natural Language Processing for Teaching and Assessment

NLP allows AI systems to comprehend, analyse and generate human language and it therefore has special relevance for teaching and assessment activities. In educational contexts, NLP is extensively used for automated grading of written responses, findings of any plagiarism and generation of feedback (Ke, 2023).

Conversational AI systems, such as chatbots and virtual teaching assistants, conversational AI systems make use of NLP, in order to assist with student queries, explanations and interactive learning. NLP techniques are also helpful in analysing posts in a discussion forum as well as reflective writing to determine learner engagement and conceptual understanding. Despite all its possibilities, NLP-based evaluation is also faced with challenges related to semantic understanding, bias and contextual interpretation. Keeping language-based evaluation fair and accurate is a key issue, particularly in learning settings that are diverse and multilingual in nature.

2.7.3 Computer Vision for Engagement and Behaviour Analysis

CV technologies support the analysis of visual information such as images and videos to derive information on learner engagement and behavioural patterns. In educational contexts, CV is being implemented to track interactions within a classroom environment, fairly identify facial expressions and process posture or gaze in order to estimate attention levels. These applications are especially pertinent in the online and hybrid learning environment in which traditional cues of student engagement may be limited. CV-based systems can deliver real-time feedback to educators and help to support adaptive instructional strategies.

However, the fact that CV is being used in education presents us with significant ethical and privacy concerns. Issues concerning consent, surveillance and the misuse of data need to be carefully addressed in order to ensure that vision-based AI systems are used responsibly.

As potted in Table 2.3, various AI technologies meet different needs in the educational domain, it is thus important to choose the proper techniques according to pedagogical goals and ethical considerations.

TABLE 2.3
AI Techniques and Their Educational Applications

AI Technique	Primary Educational Application	Example Use Cases	Key Benefits	Key Limitations
Machine learning	Learning analytics	Performance prediction, dropout detection	Scalable data analysis	Limited interpretability
Deep learning	Adaptive learning systems	Personalised content delivery	High predictive accuracy	High data requirements
Natural language processing	Teaching and assessment	Automated grading, chatbots	Efficient feedback	Bias and context sensitivity
Computer vision	Engagement analysis	Attention and behaviour monitoring	Real-time insights	Privacy concerns

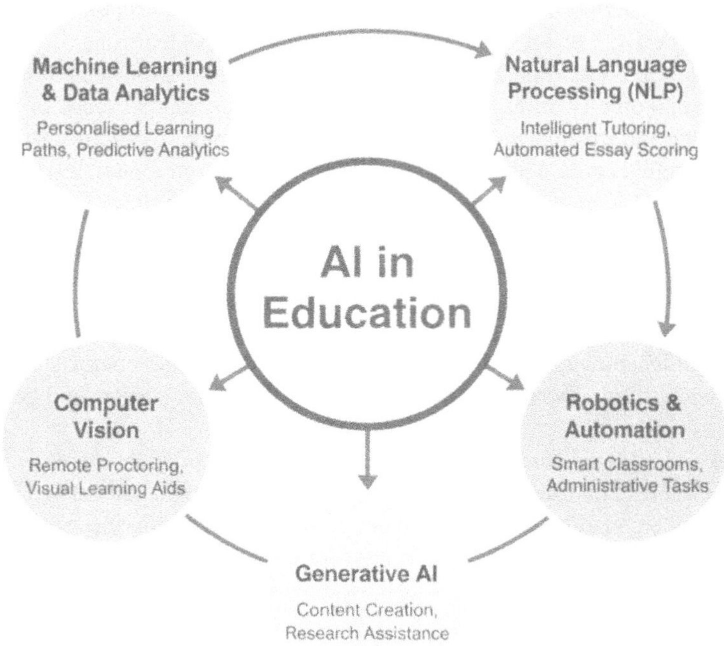

FIGURE 2.1 Core AI technologies and their role in education.

As seen in Figure 2.1, ML, natural language processing and CV working together can provide a comprehensive framework of AI for teaching, learning and assessment in education.

2.8 AI APPLICATIONS IN TEACHING, LEARNING AND ASSESSMENT

AI applications in education have the greatest impact if they directly aid in the effectiveness of teaching and learning and in learner engagement as well as equitable assessment practices. Rather than being standalone technologies, the use of artificial intelligence systems is now more common as assistive technologies that augment instructional design concepts, for customised learning and both student and educator feedback in an encouraging timely manner. This section discusses the major areas of application of AI in teaching, learning and assessment, with a focus on the practical application use cases and the realities of implementation (Long & Magerko, 2020).

2.8.1 Intelligent Tutoring Systems and Adaptive Learning

ITS are among the oldest and most mature examples of the applications of AI in education. Modern ITS make use of ML and data-driven learner modelling to deliver personalised learning by providing adaptive instruction to individual needs. These systems analyse responses from the students, the speed of the learning as well as error patterns and adjust the difficulty of the content and the selected instructional approach accordingly.

Adaptive learning platforms build upon this concept by dynamically and continuously revising leader profiles using information from real time interactions. By actually identifying gaps of knowledge and learning preferences, AI driven adaptive systems aid differentiated instruction at scale. Empirical studies are suggesting that such systems can help better engage and retain learners when embedded well in terms of curriculum design.

However, at the same time, the efficacy of ITS and the adaptive learning hinges to a large extent on pedagogical fit as well as on teacher involvement. Left to itself, with no clear instructional aim and minimal human supervision, adaptive systems are likely optimised for short-term improvement (as opposed to deep conceptual understanding) (Mandinach & Schildkamp, 2021).

2.8.2 Automated Assessment and Feedback Systems

One area that AI has received considerable approval is in assessment, especially where learning is done on a large scale and/or online. Automated assessment systems refer to systems which are designed to use AI techniques to assess objective responses, short responses and more recently, written open-ended responses.

These systems enable rapid feedback, which reduces the burden of grading, and gives greater common formative assessment. With the feedback US-relevant recommendations for learning resources this could have a biggest impact on teaching and learning. Such feedback that is given in time is linked with better learning outcomes in conjunction with the instructor's guidance.

In spite of all these advantages, automated assessment systems lack no problems of fairness, transparency and understanding the context. Over-reliance on automated assessment grading without human moderation may perpetuate or entrench bias or may misinterpret student intent, and so depends heavily on human moderation, supporting the importance of hybrid grading models.

2.8.3 Personalised Learning Pathways and Content Recommendation

AI-powered recommendation systems, which use learner data to provide personalised learning recommendations for learners regarding which learning paths, resources and activities they should pursue. These systems attempt to match the instructional content to the explicit goals and abilities of individual learners and the progress achieved by them.

Content recommendation engines are common across digital learning platforms to recommend content to students effectively navigating through complex curricula. By adjusting learning sequences and pacing, AI-based personalisation to learners can boost the motivation of learners and minimise cognitive overload. However, issues of learner autonomy and algorithmic bias are introduced by individualisation. Excessive reliance on automated recommendations could be a limitation in terms of exposure to diverse views or limitation of learning trajectories. This brings up the issue for effective personalisation to have, then, transparency, agency for learners and oversight by educators.

2.8.4 AI-Supported Classroom Management

AI applications are also increasingly being applied in domains apart from instruction and evaluation in the area of classroom management and administrative support. Predictive analytics tools allow educators to identify vulnerable students and proactively provide help to them. In a hybrid and online environment, AI systems are potentially able to help the monitoring of participation and engagement patterns.

AI-supported classroom management tools could even be used for simplifying administrative tasks like attendance, schedules and resources, etc. By taking the administration out of their hands, these applications give educators more time to teach and interact with students. Nevertheless, there is a need to be cautious in the use of AI in monitoring the classroom. Ethical considerations encompassing issues of surveillance, consent and data privacy are especially pertinent and underscore the importance of providing proper governance frameworks and implementation.

Taken together, the applications mentioned in this section reveal the potential for how AI can positively affect teaching, learning and assessment when implemented as a supportive and human-centred learning tool as opposed to being an autonomous substitute for educators.

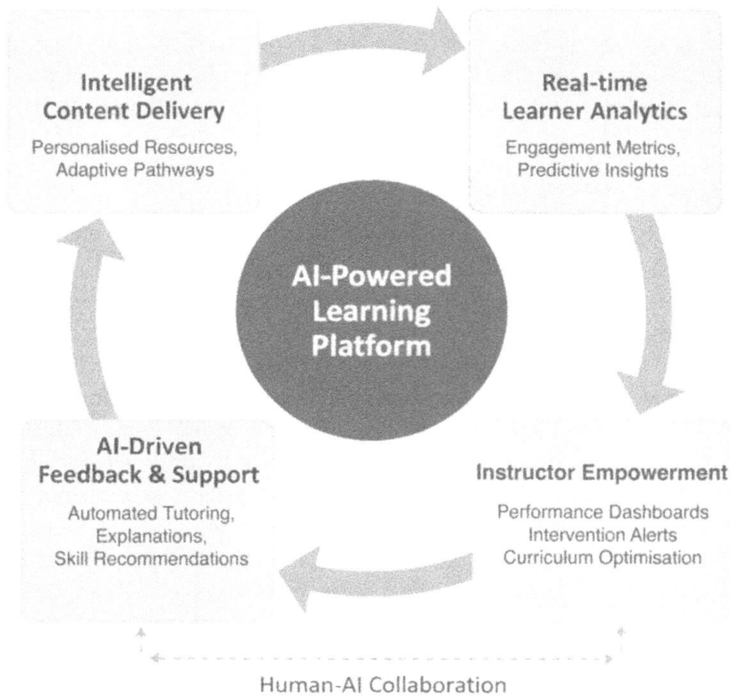

FIGURE 2.2 AI-enabled teaching-learning interaction model.

As demonstrated in Figure 2.2, AI-enabled teaching-learning interactions foster the human-AI collaboration model: Intelligent systems help educators, but pedagogical judgement cannot be replaced by AI.

2.9 LEARNING ANALYTICS AND STUDENT PERFORMANCE PREDICTION

Learning analytics has become one of the most useful and effective use cases of AI in education. By systematically analysing the data generated by the learners, AI-driven analytics can help educators and educational institutions to monitor the progress of the students, predict academic outcomes and design timely interventions. When developed and implemented responsibly, learning analytics can be used to support personalised instruction and reduce drop-out rates and support institutional decision-making. This section explores some of the major elements of learning analytics, including a major focus on student monitoring, early warning systems and decision support.

2.9.1 Data-Driven Student Monitoring Systems

Data-driven student monitoring systems gather and analyse various data sources in education such as assessment scores and attendance data, but also interaction logs, and learning management system activity. ML models are used to analyse these data and find patterns with respect to engagement, performance and learning behaviour patterns. Such systems allow students to be continually monitored instead of only by periodic assessments. Dashboards And visual analytics tools enable to translate the narratives of complex data outputs into insights for educators that can be then turned into actions – this can help educators identify students who might need additional support. Importantly, or rather, effective monitoring systems are designed to complement teacher judgement and not become a substitute for teacher judgement. Based on this potential, student monitoring systems rely very much on data quality and context interpretation. Without a pedagogical context, analysis results could lead to a misinterpretation or oversimplification of student behaviour.

2.9.2 Early Warning Systems for Dropout and Learning Gaps

Through analysing the historical data and also the real-time data, EWS is able to produce risk scores or alerts that activate timely intervention. Research has shown that early diagnosis of the learning gap will result in specific academic intervention, mentoring and counselling to really improve retention and success rate. However, the effectiveness of EWS depends on the operation of alerts in institutional processes. Automated predictions without follow-up human assistance hardly ever produce meaningful results in improved situations. Ethical issues are also at the heart of early warning systems. Transparency and fairness, and not stigmatisation are key items to ensure that the predictive insights are being used to support the students rather than labelling them negatively (UNESCO, 2021).

2.9.3 Institutional Decision Support Using AI

Beyond the course-level learners, the rapidly increasing use of AI-powered learning analytics is taking a toll at an institutional level. Universities and educational organisations are using analytics to help them design their curriculum, allocate funding, plan for enrolment and create policies.

The commonly employed decision support systems use data (aggregated learner data and learning course and programme data part) to spot systemic trends and bottlenecks in system performance (AI-based decision support systems). Such insights can be used to inform evidence-based reform and strategic planning. Used in conjunction with institutional objectives and governance structures, decision support systems that are driven by analytics increase institutional accountability and efficiency. Nevertheless, come to institutional decision-making, there are some grave concerns related to data governance, bias and accountability in AI's use. Decision support systems will therefore need to be applied with clear control mechanisms and in accordance with educational values.

As summarised in Table 2.4, learning analytics indicators based on diverse educational data sources are helpful for predictive, preventive and strategic educational outcomes when coupled with appropriate AI techniques.

As shown in Figure 2.3, AI-based learning analytics promote the systematic monitoring of students as well as the early prediction of student performance to enable timely academic interventions.

TABLE 2.4
Learning Analytics Indicators and Educational Outcomes

Analytics Indicator	Data Source	AI Technique Used	Educational Outcome Supported
Assessment performance	Exams, quizzes	Regression, classification	Academic achievement prediction
Engagement metrics	LMS interaction logs	Clustering, pattern mining	Identification of disengaged students
Attendance patterns	Institutional records	Time-series analysis	Dropout risk detection
Learning progression	Activity sequences	Deep learning	Personalised learning support
Institutional trends	Aggregated learner data	Predictive analytics	Policy and planning decisions

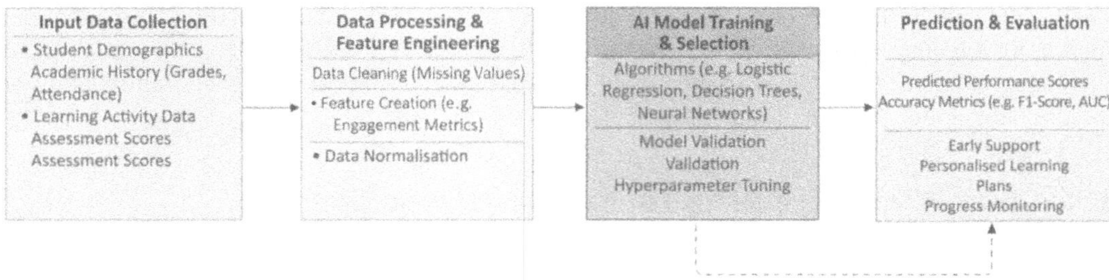

FIGURE 2.3 AI-based student performance prediction workflow.

2.10 MYTHS, MISCONCEPTIONS AND OVERHYPED CLAIMS OF AI IN EDUCATION

As AI is constantly in the pneumatic rattles in educational discourse, often exaggerated claims and misconceptions about its abilities and impact. These narratives are often driven more by marketing strategies than empirical evidence, and influences in the media and techno-optimistic assumptions. While AI provides useful tools for aiding education, blind acceptance of claims like these can create unrealistic expectations, inappropriate implementation and pushback from educators. This section discusses some of the most common myths about AI in education and explores how they compare to reality.

2.10.1 "AI Will Replace Teachers" – Myth vs Reality

One of the most stubborn myths is the one where IA is expected to replace teachers. This claim is often supported by the progress made in automated assessment, ITS and conversational agents. In reality, when educational research into this matter has been performed, it is consistent in proving the irreplaceable role teachers play in providing learning, socio-emotional support and contextualised Seventh Grade Math online school instruction. AI systems are probably best defined as assistive technologies gleaning from teacher abilities and not replacing them. Intelligent tools can be used to support lesson planning, formative assessment and learner monitoring learning, but it has a reality, they do not contain the pedagogical judgement and empathy and real and physical environment needed to teach effectively. Empirical evidence seems to support the fact that the most successful implementation of AI has taken place in contexts where teachers still play a central role in making instructional decisions.

2.10.2 Fully Automated Learning: Practical Limitations

Another common misconception is the use of education in the form that can be automated completely by AI-driven platforms. Proponents of this view in particular, in order to support their argument, would often point to the idea of adaptive learning systems and personalised delivery of content, which are ways of reducing the need for human intervention. However, fully automated learning does have serious limitations in both practice and pedagogy.

Learning is a complex social and cognitive process rather than the absorption of information and performance measurement. Automated systems have a difficult time considering motivation, creativity, collaboration and ethical development. Also, technical challenges such as data bias, lack of generalisation and no-interpretation geometrically limit the effectiveness of fully automated solutions. As a consequence, a better use of AI facilitated learning platforms implies hybridisation (blended or hybrid learning models with human facilitation and automation) when human facilitation and technology automation have to go hand in hand.

TABLE 2.5
Common AI Myths vs Actual Educational Capabilities

Myth	Common Claim	Actual Capability	Evidence-Based Perspective
AI will replace teachers	Automation makes teachers obsolete	AI supports, not replaces, educators	Teacher-AI collaboration yields best outcomes
Fully automated learning is feasible	Learning can be entirely AI-driven	Human guidance remains essential	Blended learning is more effective
AI ensures unbiased decisions	Algorithms are neutral	Models may inherit data bias	Human oversight is required
AI guarantees improved learning	Technology alone improves outcomes	Pedagogy determines effectiveness	Instructional design is critical
AI adoption is universally scalable	One-size-fits-all solutions	Context matters	Local adaptation is necessary

2.10.3 Media Narratives and Unrealistic Expectations

Media representations of AI in education are more of a disruption and transformation, and AI is represented as a force with the potential to revolutionise the educational system and quickly tackle educational system-wide issues (Table 2.5). This kind of narrative adds to unrealistic expectations by policymakers, institutions and educators. This hype-driven framing has lead to the premature adoption of immature technologies, misrule of resources as result might not be delivered in as anticipated. In addition, it can dominate important discussions on ethics, equity and long-term sustainability. A more responsible narrative of AI acknowledges both AI as a technology that is supportive and numinous as well as evolving and groundbreaking – and whose impact depends on thoughtful integration, empirical validation, and alignment with educational values.

A responsible narrative around AI would acknowledge AI as a technology that can be both supportive and numinous, evolving and groundbreaking – and whose influence depends on the wise integration, empirical validation and alignment with educational values.

2.11 CASE STUDIES: EVIDENCE-BASED AI USE IN EDUCATION

While conceptual discussions and theoretical models give important insight into AI in education, real-life implementations give the most valuable evidence of AI's real impact. This section presents some of the selected case studies on how AI has been put into effectively use in classroom settings and at the institutional level. Rather than promoting idealised success stories, a practical focus is on outcome, challenges and lessons learned, which is also consistent with an evidence-based point of view (Fikri & Rhalma, 2024).

2.11.1 AI-Supported Classroom Learning

AI-aided classroom education is one of the most direct applications of AI in education. There are several cases of successful implementation where AI-powered tools have been coupled to teaching practice in order to help achieve personalised instruction, formative assessment and student engagement.

For instance, adaptive learning platforms have been applied in the secondary and higher education classroom to personalise the course material according to individual learner progress. These systems analyse responses to assessment and interaction data in order to recommend specific exercises and resources. Evidence from such deployments suggests evidence of increased engagement of the learner and better alignment in terms of timing and pacing of instruction with student needs.

Importantly, successful teachers that implement these in their classrooms emphasise teacher control and oversight. Educators are still in charge of instructional decisions, while AI tools are employed as decision support. This hybrid approach has been proven to lead to trust between teachers and students, which decreases the resistance to technology adoption.

2.11.2 Institutional-Level AI Deployment

Beyond single classrooms, several educational institutions have implemented AI systems on an organisational level to aid academic management and student retention as well as strategic planning. Common applications include dashboards for their learning analytics, early warning systems, and also their automated administrative support.

One of the most widely reported use cases in the institutional use of AI deals with early warning systems that were developed by AI to flag out at-risk students. By using attendance data, assessment and engagement indicators, institutions were able to target earlier and provide academic advising and support services. Evaluations of such systems reflect improvement in retention rates at a measurable degree when predictions are coupled with human-led intervention strategy.

However, institutional deployments are also plagued by problems of data governance, model transparency and scalability. Institutions who have put effort into clear governance frameworks and training of stakeholders had more sustainable outcomes compared to institutions who adopted AI tools in a purely technology-driven way.

2.11.3 Lessons Learned from Real Implementations

Across classroom-level and institutional-level case studies, there are a number of recurring lessons. First, the existence of AI is also context dependent; that is, the tools have to be adapted to local pedagogical, cultural and infrastructural contexts. Second, human involvement (especially that of educators and academic advisors) continues to play an important role in converting AI insights into fruitful educational action.

Third, transparency and communication play an important role in trust building. Students and educators are more likely to use an AI system when the purpose of its use, its limitations and how any data are used have been clearly explained. Finally, long-term success is a matter of perpetual evaluation and improvement, not one example. These lessons are instrumental to this chapter's main argument: AI in education is valuable not only as a tool for automation and assessment – it is beneficial as a responsible and substantiated integration prioritising learning outcomes and human judgement. The case studies covered in this section illustrate evidence of how building on evidence of what works in human-centred deployment of AI can improve educational practice while avoiding the pitfalls of hype-driven adoption of these (Fikri & Rhalma, 2025).

As illustrated in Figures 2.4 and 2.5, instance-based applications of AI in the educational context led to positive measurable improvements of learning supporting, provided it is coupled to institutional and content-related learning.

2.12 FUTURE DIRECTIONS AND RESPONSIBLE AI ADOPTION IN EDUCATION

With artificial intelligence reveals of entire evolution its place in education will likely continue developing from current purposes than more transparent, inclusive and ethically grounded systems. Future advancement of AI-enabled learning will depend not only on innovations in technology but also on responsible ways of adoption with pedagogical values, social equity and institutional readiness. This section highlights some of the future directions and road map of AI in education, with the focus on trustworthiness, inclusivity and governance.

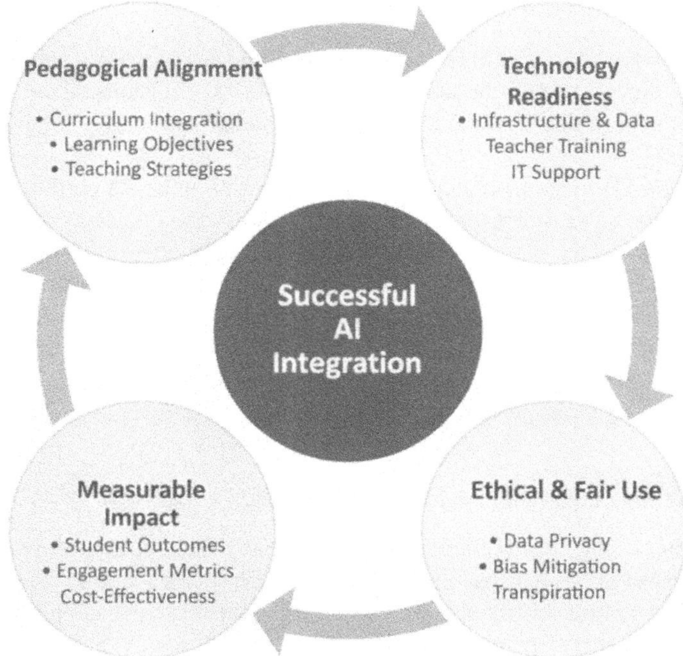

FIGURE 2.4 Evidence-based AI adoption in real educational settings.

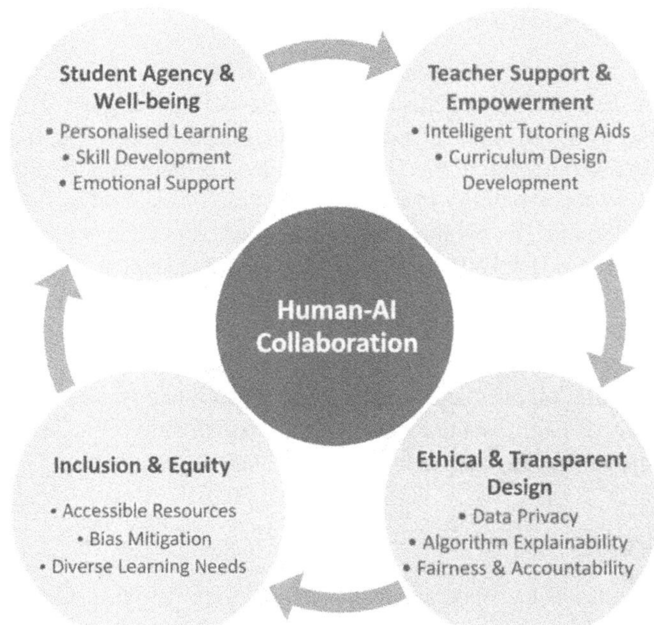

FIGURE 2.5 Human-centred AI model in education.

2.12.1 Explainable and Trustworthy AI for Education

One of the most important domains of future directions of AI is explainable and trusted AI systems. As AI increasingly plays a role in the making of instructional decisions, assessments and student support mechanisms, desirable transparency into how models make their predictions and recommendations is critical.

Explainable AI (XAI) methods understandably attempt to make AI outputs understandable to educators and learners, to the point the users can understand why decisions were made in the first place. In the context of education, explainability aids in building trust, informed intervention and enabling the stakeholders to challenge or calibrate an AI-driven insight. Trustworthy AI should also be robust, reliable and consistent when learning in a variety of environments. Yet, the predictive AI systems that will be developed in education systems of the future are predicted to incorporate explainability as a central tenet to the system's design, rather than at the confluence, in order to ensure that the intelligent tools are held accountable and aligned with the goals of education.

2.12.2 Inclusive and Equitable AI-Driven Learning

If the focus is on the setting, how realistic are your students? Without a careful design decision, AI systems risk to increase existing disparities with regard to on socioeconomic grounds, language and access to technology. Future AI-based learning environments must focus on inclusive design and equability of outcome is therefore necessary the concept of Inclusive AI is inclusive and diverse use of data sets and a pedagogical approach and adaptive systems which meets a variety of learning needs.

AI has the potential to support learners with disabilities, to enable multilingual education and to make more personalised learning opportunities available – however, if those considerations of equity are not made from the get-go. Assuring equitable AI-driven learning also calls for continued evaluation of the impact of systems on different groups of learners, which again reinforces the importance of fairness audits and participatory design approaches.

2.12.3 Policy and Institutional Readiness

Responsible artificial intelligence adoption in the education sector is dependent on institutional and policy readiness. Educational organisations need to develop aligned strategies for the long-term implementation of AI, governance frameworks and capacity building initiatives to facilitate the process of sustainable integration of AI. This includes the definition of ethical guidelines, data governance policies and accountability mechanisms. Institutions should be investing in infrastructure, professional development, and without engaging stakeholders one is to ensure that AI systems are used effectively and responsibly. Policymakers have a critical role to play in setting parameters and regulations to balance innovation for the provision of learning and ensuring the protection and trust of learners and the public. Future success for education using AI, therefore, will be defined by collaborative work between educators, institutions, policy makers and developers of technology, and emphasising collaboration over separating the use of technology. The future developments of AI-enabled education will pass from a combination of explainability, inclusivity and institutional preparedness in order to avoid any responsibilities and make it sustainable.

2.13 CONCLUSIONS

This chapter has made use of a critical and scholarly look into the place that artificial intelligence has in education by way of our journey from the hype and highly sensationalised stories to something a little bit more grounded and human-centred. While the potential of various AI technologies to assist in aspects of teaching, learning, assessment and institutional decision-making is significant, the educational value of these technologies is contingent on their responsible design, thoughtful implementation and ongoing engagement of people.

By tracing the development of AI in education from its origin and analysing key technologies such as ML, natural language processing and CV, the chapter has shown how AI is not a stand-alone solution to the challenges posed by education. Instead, its effectiveness is conditioned from pedagogical alignment, quality of data as well as in classroom and in institutions. Discussion of learning analytics, and predicting student performance disclosing the potential of AI to help us derive actionable insights by going beyond the role of being an autopilot, but using AI as a decision support tool. The case studies described in this chapter provided more evidence for the need for evidence-based adoption. Successful implementations always created a classroom environment where teachers were engaged in the process, institutional governance and transparency of the process; while hype driven deployments often failed to create meaningful learning outcomes. These results indicate the importance of disavow of the dogma combining the promise of technology with actual application in educational technology endeavours based on AI.

Ethical, privacy and governance considerations was proved to be one of the key aspects of sustainable adoption of AI. Issues around data protection, algorithmic bias and accountability can't be relegated to a secondary place, particularly as in the case of education and other areas with a vulnerable population. A human-centred approach (one that places educators and learners at the heart of education coupled with the emphasis on institutional responsibility) to the problem-solving that is done throughout the chapter. Looking into the future an important message highlighted in this chapter of this book was that the days in which AI functions in education will be explainable, inclusive and trustworthy educational environments, with clear policy frameworks within appropriate institutions.

NOVELTY BOX 2: SUMMARY OF CONTRIBUTIONS

This chapter has three major contributions to the debate on AI in education:

- Critical separation of hype vs reality: Identify AI claims through a test of empirical evidence with practical classroom experience
- Evidence-based assessment of AI applications within education, demonstrating controversies, and based on members employed cases and formulas of actual implementation in education applications
- Human-centred and ethical AI Framework that places teachers and institutions at the centre of responsibility for ethical AI adoption

REFERENCES

Aldenaini, N. (2021). Trends in persuasive technologies for physical activity and sedentary behavior: A systematic review. *Article Collection on Human Aspects in Adaptive and Personalized Interactive Environments (HAAPIE)*, 7084.

Almaskari, T. H., Mohamad, E., Yahaya, S. N., & Jalil, M. F. (2021). Leadership as a driver of employees' innovation performance: The mediating effect of cultural diversity in UAE universities. *The Journal of Asian Finance, Economics and Business*, *8*(8), 271–285.

Chassignol, M., Khoroshavin, A., Klimova, A., & Bilyatdinova, A. (2018). Artificial intelligence trends in education: A narrative overview. *Procedia Computer Science*, *136*, 16–24.

Chen, L., Chen, P., & Lin, Z. (2020). Artificial intelligence in education: A review. *IEEE Access*, *8*, 75264–75278.

Douglass, S., Scott, J. T., & Anderson, G. L. (2024). *The politics of education policy in an era of inequality: Possibilities for democratic schooling*. London: Routledge.

Dwivedi, Y. K., Hughes, L., Ismagilova, E., Aarts, G., Coombs, C., Crick, T., … & Williams, M. D. (2021). Artificial Intelligence (AI): Multidisciplinary perspectives on emerging challenges, opportunities, and agenda for research, practice and policy. *International Journal of Information Management*, *57*, 101994.

Fikri, Y., & Rhalma, M. (2024). Artificial Intelligence (AI) in early childhood education (ECE): Do effects and interactions matter. *International Journal of Religion: Politics, Sociology, Culture*, *5*(11), 7536–7545. https://doi.org/10.61707/y74fv875

Fikri, Y., & Rhalma, M. (2025). Nexus between early childhood education, artificial intelligence, and education for sustainability. *UKR Journal of Economics, Business and Management (UKRJEBM)*, *1*(3), 78–83. https://doi.org/10.5281/zenodo.16493120

Han, S., Zhang, Z., Liu, H., Kong, W., Xue, Z., Cao, T., & Shu, J. (2024). From engagement to performance: The role of effort regulation in higher education online learning. *Interactive Learning Environments*, *32*(10), 6607–6627.

Holstein, K., McLaren, B. M., & Aleven, V. (2019). Designing for complementarity: Teacher and student needs for orchestration support in AI-enhanced classrooms. In *International conference on artificial intelligence in education* (pp. 157–171). Cham: Springer International Publishing.

Ke, M. F. (2023). Applications and challenges of artificial intelligence in the future of art education. *Pacific International Journal*, *6*(3), 61–65.

Long, D., & Magerko, B. (2020). What is AI literacy? Competencies and design considerations. In *Proceedings of the 2020 CHI conference on human factors in computing systems* (pp. 1–16).

Mandinach, E. B., & Schildkamp, K. (2021). Misconceptions about data-based decision making in education: An exploration of the literature. *Studies in Educational Evaluation*, *69*, 100842.

UNESCO. (2021). *AI and education: Guidance for policy-makers*. Paris: UNESCO.

3 The Smart Classroom and IoT

Revati Ramrao Rautrao

3.1 INTRODUCTION

For the better part of two centuries, the classroom was functionally frozen. The agrarian one-room schoolhouse and the late industrial age lecture hall had a basic framework in common with a single vector of knowledge given out to passively receive recipients (Al-Khresheh & Alkursheh, 2024). Technology, when introduced, came in the form of a prosthetic appendage – the chalkboard was an extension of the teacher's arm, the projector his or her voice, the laptop digitised notetaking. These were instruments of amplification and not transformation (Alzahrani & Alhalafawy, 2023).

The modern smart classroom movement, intensified through the development of global policy (such as the Education 5.0 initiatives and national schooling digital education strategies), is an inversion of this tree (Rosenbaum et al., 2024). In the former version of technology-as-tool, it abandons the metaphor of technology-as-environment. Where the traditional infrastructures were reactive, in need of explicit human command, smart classrooms are aspiring towards proactivity; spaces that are perceptive, inferential and adaptive (Yan, 2024). Yet again, early smart classrooms were too often guilty of the same mistake that their predecessors committed: they draped technology over old pedagogy. Sensors were put in place, but they were not connected to instructional decisions (Osman & Ahmed, 2024). Data were all harvested and lying fallow in dashboards. The one thing that was missing and critical was cognateness (Xiao et al., 2025). The historical path of classroom technology may be summarised in Table 3.1, showing the transition from amplification devices to transformative smart environments.

3.2 THE AI-IoT NEXUS IN EDUCATION

In this chapter, the artificial intelligence (AI)-Internet of Things (IoT) nexus is not brought up as the combination of two technologies, but as the creation of a third phenomenon, the co-intelligent learning ecosystem. In case IoT equips the classroom with sensory organs, the ability to see, hear and touch, AI offers the interpretive cortex. Kept apart, they generate noise and lighting respectively; put together, they facilitate what this chapter refers to as ambient pedagogical agency (Knights et al., 2024). It is not just a semantic difference. The conventional educational technology presupposed clear training: the teacher clicks, the system reacts. With a co-intelligent ecosystem, the environment itself becomes a member. It is not responsive to commands; it predicts needs (Zou et al., 2025). It does not just store information; it recognises patterns that the human eye cannot see. This restructuring, not only of what transpires in the classrooms but of what classrooms are, is a shift in the instrumentality of the

TABLE 3.1
Evolution of Classroom Technology

Era	Technology Used	Role of Technology	Pedagogical Impact
Agrarian/Industrial	Chalkboard, lecture hall	Amplification of the teacher	Passive learning
Late 20th century	Projector, laptop	Digitised notetaking	Limited interactivity
Early smart classrooms	Sensors, dashboards	Data collection only	Technology draped over old pedagogy
Modern smart classrooms	AI + IoT integration	Proactive, adaptive partner	Emergent pedagogy, personalised learning

DOI: 10.1201/9781003767855-3

TABLE 3.2
AI-IoT Nexus in Smart Classrooms

Component	Function	Example
IoT	Sensory organs (see, hear, touch)	CO_2 sensors, acoustic spectrometers
AI	Interpretive cortex (analysis)	Pattern recognition, predictive analytics
Synergy	Co-intelligent ecosystem	Ambient pedagogical agency, real-time adaptation

reactive to the proactive partner (Adewojo & Dunmade, 2024). Three main points are made in this chapter. To begin with, combining IoT and AI is qualitatively different from other past technological convergences, as it provides the ability to adapt in real-time and in closed loops between the physical space and the digital systems, and human learners (Fidalgo & Thormann, 2024). Second, such convergence requires a bit of a renegotiation of the professional identity of teachers, as knowledge deliverers or pedagogical designers. Third, smart classroom ethical architecture should be prospective and not an afterthought (Ghosh & Ravichandran, 2024). The AI-IoT nexus thinking leads to a co-intelligent ecosystem, as shown in Table 3.2, whereby IoT provides senses, and AI provides interpretive capacity.

3.3 METHODOLOGY

3.3.1 Research Design

This is suitable in less developed interdisciplinary areas where the literature is heterogeneous and still not susceptible to systematic meta-analysis. This was aimed at mapping major concepts, finding clusters of evidence and gaps, as well as synthesising conceptual frameworks.

3.3.1.1 Research Questions

What are the infrastructural, pedagogical, and ethical aspects of the existing applications of AI-IoT in smart classes?

What is the operational form of the synergy between the IoT sensing and AI analytics to teach and learn?

What changes in the role of educators, development of professionalism, and assessment strategies are reported?

3.3.1.2 Search Strategy

Six databases were used: Web of Science, Scopus, IEEE Xplore, ACM Digital Library, ERIC, and Education Source. Search strings were the aggregation of terms in three blocks, including (IoT/environment), (AI/analytics) and (educational context). The literature search was conducted from January 2020 to December 2025. Database searches were supported by citation chasing. OECD, UNESCO, EDUCAUSE and ISTE were the sources of targeted grey literature.

3.3.1.3 Eligibility Criteria

Inclusion: (a) empirical, conceptual, or critical review studies on AI/IoT in physical/hybrid classrooms; (b) English language; (c) any educational level; and (d) 2020–2025.

Exclusion: only theoretical papers that lack an educational relationship; online environments fully online and lack a physical component; editorials; and duplicates.

3.3.1.4 Selection Process

Two reviewers screened titles and abstracts, with any conflict being sorted out by a third. Wholesome texts were evaluated on the grounds. Final corpus: 147 sources (112 journal articles, 21 conference proceedings, 14 grey literature reports).

3.3.1.5 Data Extraction and Synthesis

A standardised form extracted: authorship, year, country, educational level, research design, IoT/AI technologies, integration architecture, outcomes, challenges, ethics. NVivo-based thematic synthesis was used to produce 53 descriptive codes that were clustered into 8 analytical themes consistent with the chapter organisation.

3.3.1.6 Quality Appraisal

Appraisal of the empirical studies was done through MMAT; grey literature through AACODS. No research was screened out based on quality considerations; advantages and disadvantages were used to synthesise.

3.3.1.7 Limitations

Due to the rapid change in technology, innovations may only be reflected in preprints; the use of the English language might not represent non-English innovations well; pilot studies might be predominant and restrict the generalisation. Synthesis is concerned with conceptual patterns, not with general assertions.

3.4 SYNTHESIS AND CHAPTER CONSTRUCTION

The results were interwoven as a narration in the chapter. All the subsections combine empirical evidence, theoretical framing, and critical analysis. The conflicting evidence is accurately addressed. This approach will guarantee a clear, reproducible, and justifiable body of evidence.

3.5 THE INTERNET OF THINGS (IoT) IN SMART CLASSROOMS

3.5.1 IoT Infrastructure as Invisible Pedagogy

The modern smart classrooms are fabricated based on what can be called a pedagogical thread, a combination of sensors and actuators so perfectly linked to the human mind that they are no longer consciously perceived (Chiradeja & Yoomak, 2023). A full-fledged IoT classroom uses a number of different types of environmental sensors: illuminance sensors, thermohydrometers, CO_2 sensors, particle monitors, acoustic spectrometers, and passive infrared occupancy arrays (Baena-Navarro et al., 2025). This is not instrumentation as an end in itself, and every piece of data are associated with a validated predictor of cognitive performance. Topological architecture over sensor proliferation is the new techno-infrastructural innovation of recent years. Conventional deployments were characterised by vertical fragmentation to an extent that lighting was handled by a different system, HVAC by another, audiovisual by the third (Chauhan, 2024). Convergent smart classroom topology is a topology that brings together all the end devices, which include sensors in the environment, lighting, HVAC, audiovisual equipment, and power management, and centralises them to an IoT gateway (Seraj et al., 2024). It is precisely this architectural flattening that facilitates what network theorists refer to as cross-modal correlation: the ability to realise that an increase in CO_2 and fidgeting by students, which are measured by air quality sensors and AI-vision systems, respectively, are not two distinct issues but rather one phenomenon that needs to be addressed jointly (Das et al., 2025). However, infrastructure carries intangible curricula.

Automating the response to student presence in classrooms implicitly educates that surveillance is a condition of service. In the case when the environmental optimisation is done in the unconsciousness of students, it represents the world in which the agency belongs to systems, rather than citizens (Eden et al., 2024). Physical architecture of the IoT is moral at the same time, the privacy concerns cannot be limited to the policy addenda but must be cut into system logic at the firmware level (Antonijevic et al., 2025).

3.5.2 From Environmental Sensing to Atmospheric Intelligence

Reconsidering classroom atmosphere not as a background to but as a variable is needed to make the switch between the non-adaptive and adaptive learning environment (Kamil & Abdulazeez, 2024). The recent extensive research examining tens of thousands of environmental data points in a wide range of teaching practices, lectures, discussions, and independent study has shown that environmental sensitivity depends on the scenario (Al Mamun & Lawrie, 2023). Engagement in independent working is best predicted by temperature, and in lectures and discussions, the illuminance and CO_2 saturation have stronger influences. This observation breaks down the belief of an overall, so-called ideal classroom climate (Ragab et al., 2025). Consequences are more than comforting. The insights that these models have brought out are the fact that the physical environment is not neutral, but pedagogical. CO_2 levels are not only uncomfortable but also degrade executive functioning and lower the same abilities of critical analysis and sustained attention that learning is supposed to develop (Gravett & Lygo-Baker, 2024).

Not only is a classroom with poor ventilation unpleasant, but it is incompetent in pedagogical terms. The next horizon is the atmospheric intelligence: systems that do not simply perform the predetermined environmental profiles but assume the suitable conditions based on the instructional situation (Kim et al., 2025). An AI-vision system would understand when a lecture changes into small-group work, and the environment does not wait to be manually selected; it anticipates the change and responds to it by switching the lighting zones and acoustic damping (Sajadieh & Noh, 2025). Its new implementations use digital twin interfaces that are virtual versions of real-life classrooms, by providing teachers with physical environmental settings to monitor and control. The adult form gets rid of the interface, and the command is substituted by consent.

3.5.3 Wearable Technology and Biometric Data in Education

Wearable technologies in education until now have been either a dystopian caricature or a pilot project of irrelevance (Sudan et al., 2024). This is changing with the non-invasive physiological sensing. Heart rate variability can now be sensed by photoplethysmography within a wrist-worn form; electrostatic response can be measured by electrodermal activity sensors, which serve as the index of the body's sympathetic nervous system response; the body's attention state, oculomotor tracking can be estimated using optical sensors (Alshanskaia et al., 2024). The pedagogical promise is genuine. Moment-to-moment autonomic data can provide hints of cognitive mismatches between the demands involved and the responsibilities. Gaze forward, posture stay still – a student looking attentive may be under a state of effortful struggle or effortful disengagement. Physiology is the key to distinguishing (Gkintoni et al., 2025). Early applications in adaptive learning systems demonstrate that incorporating galvanic skin response into content sequencing algorithms reduces self-reported frustration and increases persistence on challenging tasks (Rajput, 2025).

Yet the ethical environment is unprecedented. There are some fundamental differences, such as biometric data and behavioural data. A student can control his or her behaviour in class; a student cannot control his or her electrodermal response. When educational systems get access to involuntary physiological signals, they cross a boundary which it was separating education from medicine before (Kłęczek et al., 2024). Recent studies of trust in sensor-learning environments show a complex trust calculus for students, who differentiate data and methods collected for real-time adaptation and data methods stored for predictive modelling, and are more comfortable with the former (Dudek et al., 2025). It is therefore possible that there is a certain design principle here: it is temporal proximity that acts as a mediator of the legitimacy of privacy. Systems that take immediate and decisive action based on sensed data, and then discard it, may generate trust that is unattainable for perpetual surveillance systems (Koukaras et al., 2025). These ethical issues are summarised in Table 3.3, which describes privacy, biometric boundaries, predictive analytics and trust issues.

TABLE 3.3
Ethical Considerations in Smart Classrooms

Area	Concern	Example
Privacy	Surveillance embedded in infrastructure	Automated attendance via sensors
Biometric data	Involuntary physiological signals	Electrodermal activity tracking
Predictive analytics	Algorithmic streaming & bias	Sorting students into trajectories
Trust & legitimacy	Temporal proximity of data use	Real-time adaptation vs. perpetual storage

3.6 ARTIFICIAL INTELLIGENCE (AI) IN SMART CLASSROOMS

3.6.1 AI-Driven Data Analysis and Insights

The amount of data produced by instrumented classrooms is of an order of magnitude greater than human cognitive capacity. A single sensor-heavy classroom will generate thousands of environmental data points per hour, compounded by AI-vision analytics on behavioural inferences per second, constituting numerous orders of magnitude of bountiful data flowing to the classrooms that rely on these technologies (Dudek et al., 2025). Raw data of this scale is no insight; it is noise. The part played by AI in this ecology is transmutation. Deep learning architectures in general and Long Short-Term Memory networks and attention-based transformers in particular – are good at finding out the dependencies over time that cannot be recognised by linear models (Sasal et al., 2022).

The contemporary frameworks of educational AI show that student interaction sequences, modelled as high-dimensional embeddings, encode predictive information both relating to engagement right away as well as relating to learning trajectories over time. What may appear to a human being as random variation in help-seeking behaviour may represent to a neural network, working with millions of other similar sequences, to be recognised as the prodromal signature of conceptual misconception (Fazil & Albahlal, 2025). This capacity gives rise to a new visibility of education. Teachers have always known that some students struggle, but with AI, one can know when they begin struggling, what modality of struggling they are exhibiting and what intervention has been historically efficacious for a learner exhibiting identical 'signatures'. Reduced latency in the delivery of content and accuracy in allocation of resources are not just network measures; they translate into being instructionally responsive content that arrives when it is necessary, at the granularity necessary (Mehta et al., 2023).

3.6.2 Adaptive Learning Systems and the Personalisation Paradox

Adaptive learning platforms are considered AI's most mature application in education; however, they are still trapped in what one could describe as the personalisation paradox. Their foundational premise – that instructional content should dynamically calibrate to learner performance is of fact empirically sound (Naseer et al., 2025). In a series of meta-analyses, it has been shown that the results of mastery-based adaptive systems lead to effect sizes larger than one standard deviation compared to tutoring interventions (Wang et al., 2023). The paradox develops in implementation. Current adaptive systems do very well when it comes to optimising the acquisition of well-structured knowledge: mathematics procedures, vocabulary, and factual recall. They are not particularly good at devising ill-structured knowledge: historical interpretation, scientific argumentation, and literary analysis (Maity & Deroy, 2024). This is not a temporary, limited technical aspect but an underlying around structure limited. Adaptive algorithms need proper responses to be clear for them to function; they need problems with one correct answer.

The loftiest dreams of education are about precisely those things for which there is no one correct response (Alexsius Pardosi et al., 2024). This chapter suggests conceptual reframing. Instead of personalisation as content-matching, we can, perhaps, think of it as challenge-configuration (Imran et al., 2024). The function of the AI's role is not to decide what problem the student should solve but to put together the conditions, the resources, collaborators, time constraints, and representational formats, and the student solves problems of their own choosing (Saputra et al., 2024). This change from somewhat algorithmic prescription to algorithmic scaffolding is consistent with the new generative AI functionality supporting open-ended inquiry as opposed to closed-loop instruction.

3.6.3 Predictive Analytics and Intervention Strategies

Educational prediction has always been based on annual cycles: one's performance on a standardised test in spring predicts one's placement next autumn (Ajuwon et al., 2024). AI makes it possible to compress this temporal rather large-scale to minutes and hours. Recent implementations have shown network-level quality of service degradation to have high accuracy in predicting disengagement in students, about 90 seconds before disengagement behavioural manifests. What this is one year away, predicting the drop is not a prediction of next year's dumb drop is a prediction of the next minute's withdrawal. The ethical valence of such capability is radically ambiguous. In one register, it makes it possible to introduce pedagogic care in a preemptive manner – before frustration is consolidated into disidentification (Mahmood et al., 2024). The teacher who gets an alert that bandwidth insufficiency is failing a remote student's video stream can intervene to fix any connectivity before a student gives up the lesson (Naciri et al., 2021). In another register, it enables what critical theorists call the algorithmic streaming (i.e. sorting students into trajectories of expectation that are self-fulfilling prophecies).

The regulatory landscape remains nascent. Existing data protection frameworks do approach substantive protections as they provide procedural guardrails but fail to offer remedies on the substantive question of whether or not educational systems legitimate the acquisition of predictive models about individual students (Wu & Yusof, 2024). This chapter argues that the critical view of the ethical boundary is not the quality of predictive accuracy, but the predictive use. Models used to allocate instructional support area in a different moral register than those used to allocate educational opportunity (Thuy & Tien, 2024). The former helps: the latter forecloses.

3.7 THE SYNERGY OF AI AND IoT IN SMART CLASSROOMS

3.7.1 Real-Time Feedback and Continuous Assessment

Summative assessment is an information-poor artefact. When teachers were only able to observe students at discrete intervals, evaluation required discrete events (Krumsvik, 2025). The AI-IoT classroom helps to dissolve this constraint. Every interaction becomes assessment; every response, data; every hesitation, signal (Gill et al., 2023). Recent educational platforms represent the shift in paradigm. IoT sensors transmit real-time environmental or experimental conditions to generative AI engines, which create dynamic localised problem scenarios where learners are required to interpret current data, along with domain knowledge. Assessment is not added onto instruction; it goes together with it (Amin, 2023). The system does not test what students learned after the fact; it has an ongoing process to calibrate its estimate of their competence with each response (Li, 2024).

This compression of the assessment cycle from weeks to milliseconds allows for some pedagogical modes not available to traditional classrooms, namely micro-interventions (Luo, 2023). Rather than waiting for the results of the unit tests to find struggling students, the system detects (Vulic et al., 2024) within one problem-solving episode the exact moment at which a misconception arises. Feedback is not waiting and generalised; it is immediate and specific. Evaluations of such platforms report not only on gains of knowledge, but qualitative shifts of effect of learners – they report on feeling "talked with" rather than "tested by" the system (Naseer & Khawaja, 2025).

A layered flowchart illustrating the integration of the IoT to sensors in the physical environment feeds data into IoT AI processing systems, which in turn enable pedagogical modifications of the environment (lighting, HVAC adjustments, scaffolding instructional topics, etc.).

3.7.2 Cognitive Scaffolding and Collaborative Intelligence

The dominant metaphor of educational technology has been that of tutor – one system, one learner. AI-IoT integration allows for a different setup, the orchestrated collective (Maity & Deroy, 2024). If students' devices, every environmental sensor, and every learning object have a common interest, specific to a common data fabric, cooperative learning is no longer a logistical or homework but the enthusiasm around (Rathika et al., 2024). Consider a small-group problem-solving episode in smart classroom, where microphones record not only who speaks, but it is also possible to gauge prosodic elements of uncertainty or confidence by students (Tutul et al, 2024).

AI-vision systems monitor gaze direction and therefore inform us about the shared artefact that currently controls the attention of the group. Individual devices are used for streaming the real-time interaction with digital resources (Zhao & Wang, 2022). These streams become conflated in one representation of collective cognitive state – not just what the group has done, but how it is thinking. This capability provides for what may be called swarm scaffolding: instructional interventions that are designed not for individuals but for groups. When we understand that the discussion is getting stuck on one representational format, we may provide the group with an alternative visualisation dynamically on the display (Galesic et al., 2023). When it identifies that it has not heard vocalisation for an extended period of time after a complex prompt, it may inject a structured turn-taking protocol. The technology does not replace the collaboration of humans; on the contrary, it makes the collaboration legible and thus improvable.

3.7.3 The Physical Classroom as Co-Teacher: Ambient Pedagogical Agency

The ontological implication of AI working with IoT is the most profound of all. As environments are perceiving, inferring, and adjusting, learning is actually an actor in the classroom (Mishra & Varshney, 2024). This chapter defines a construct of ambient pedagogical agency to describe this phenomenon. Agency, as we are thinking of it here, does not mean conscious or intentional in a human sense (Wen et al., 2024). It represents causal efficacy to the instructional system. The customary classroom was very deterministic about agency-granting it solely to her or his human participants; the habitat was the stage, not the performer. In the co-intelligent classroom, environmental adaptations carry pedagogical consequences (Arifin et al., 2024; Kwarteng et al., 2025).

A lighting change which increases focus when changing from individual to group work is not simply facilitative but instructional. Implicitly, it teaches that there are different modes of attention associated with different activity structures (Sieg et al., 2025). Contemporary implementations have proven this principle to work. When CO_2 sensors detect the deterioration of the air during an extended lecture, the system does not simply turn up ventilation; rather, it synchronises this change in its environmental surroundings with instructional flow, placing ventilation change timings in the natural transition points in the instructor's discourse (Mousavi et al., 2024). The environment is not disturbed, but it choreographs. But students experience not technical intervention but seamlessness. This ambient agency raises questions that educational theory has yet to formulate, let alone answer. If your classroom teacher is responsible for the teaching of the? At the end of the day, when such environmental adaptations have differential consequences from student population to student population, is this pedagogical discrimination or system optimisation? The legal framework of education recognises teachers and institutions as responsible actors; they do not yet recognise environments. This is a frontier not only in terms of technical development, but educational jurisprudence.

3.8 THE CHANGING ROLE OF EDUCATORS

3.8.1 Teachers as Pedagogical Designers

The rhetoric of technological displacement has been an anticipation of educational discourse since Skinner's teaching machines. The testimony from cases of mature implementations of AI-IoT is more favourable to a different story: not substitution but transformation of professional practice (Ramakrishnan et al., 2024). Emerging large-scale primary school teacher research trained in IoT and data modelling with AI is helpful in establishing the ground of evidence. Following targeted professional development, teachers evidenced statistically significant improvement based on established competencies of pedagogy from a technological perspective (Lakhe Shrestha et al., 2025). More interestingly, significant majorities produce artefacts indicating digital creativity in IoT system design and have also successfully embedded AI aspects into their pedagogical designs (Lakhe Shrestha et al., 2025).

These aren't workers resisting automation; they are professionals taking new tools and using them to increase their professional agency. The emergent professional identity can be described perhaps best as a pedagogical architect. The architect does not actually fabricate all of the building's various elements but designs the system within which fabrication takes place. Similarly, it is not the smart classroom teacher's place to write the adaptive algorithm code, but to structure the learning objectives, structures and success criteria for the tasks they serve (Ramakrishnan et al., 2024). Their expertise moves from being about content delivery to experience design – from telling to curating, from assessing to interpreting. Table 3.4 shows the shifting role of educators, which has gone from being the deliverer of knowledge to the role of pedagogical architect.

This flowchart shows the professional transformations of the educator in smart classrooms. The journey commences with the Knowledge Deliverer that is content transmission-oriented. It develops into the Pedagogical Designer: the curates learning experiences with the help of AI-IoT tools. Finally, it culminates in the Architect of Learning Ecosystem, who designs adaptive systems and puts dynamic learning environments together. Each stage corresponds with a change in agency, innovativeness, and technological savvy.

3.8.2 Professional Development and Digital Literacy

Conventional technology professional development – one-time workshops devoted to the use of software – leads to what educational researchers mean when they say you have "syntactic competence": "the ability to execute procedural sequences without deep understanding" (Gong et al., 2025). Effective present programmes, however, build full frameworks for competency growth that include a theoretical basis, in-depth experimentation, evaluation advice, and certification check (Boscardin et al., 2025). Particularly innovative is the focus on what might be termed educational translation – i.e., transforming technical concepts into pedagogically operative knowledge (Long et al., 2024). Systems thinking principles borrowed from the fields of computer science are neither taught as a network architecture nor remapped as curriculum design principles: learning activities (once designed) should dynamically deploy within individual, small group, and whole class

TABLE 3.4
Changing Role of Educators

Traditional Role	Smart Classroom Role	Skills Required
Knowledge deliverer	Pedagogical designer/architect	IoT system design, AI integration
Content delivery	Experience curation	Digital literacy, systems thinking
Assessment	Interpretation & scaffolding	Adaptive learning strategies

configurations (Jetzinger et al., 2024). This process from technical register to pedagogical register is one of the differences between transformative professional development and merely informational training.

Evaluation components address professional development's perennial weakness: transfer. By not only equipping teachers to implement technology-enhanced curricula but also by building on the idea that embedding process evaluation frameworks into the training itself encourages and enables teachers to consider the quality of their implementation rather than simply containing a measure of effectiveness. This metacognitive layer – which is learning to assess one's own technological pedagogy – may make up the critical demarcation line between novice and expert practice.

3.8.3 Resistance as Rational Response

Educational technology literature has tended to pathologise the resistance of teachers, describing it as Luddism or professional insecurity. A different interpretation deserves consideration, that of resistance being rational risk assessment (Viberg et al., 2024). Teachers in AI-IoT environments are faced with actual professional vulnerabilities. Algorithmic systems to predict student performance produce visibility where there existed professional discretion. It is not easy to falsify the idea that a student is "trying hard" but struggling, an idea that makes the teacher's judgement, whereas the idea, a prediction produced by an AI algorithm that cookies is a school teacher-that the same student has a quantified probability of failing a course is numerically exact and thus professionally authoritative (Gordon et al., 2024). Resistors to the adoption of AI may be responding not to the technology but to the potential of AI to devalue their professional judgement (Sidorkin, 2025).

This chapter argues for the proposition that sustainable integration involves epistemic justice – recognition of the fact that teacher professional knowledge and algorithmic pattern recognition are not hierarchical forms of knowledge, but rather distinct ones. The former are best at holistic judgement that is sensitive to context provided by relational history; the latter excel at picking out subtle multivariate correlations among large groups of people (Lunn Brownlee et al., 2022). Neither reduces to the other. Professional development must therefore develop not only technical skills but also critical data literacy – the ability to question algorithmic outputs and how they are limited and assert professional judgement when there are disagreements between algorithmic and human judgement (Wu & Yusof, 2024).

3.9 THE FUTURE OF SMART CLASSROOMS

3.9.1 Emerging Technologies and Their Potential Impact

The curve of the integration of AI and IoT advancement is not a straight line but a qualitative change. Three new areas need special attention.

Generative AI-Physical convergence: There is already a generation of appropriate educational platforms that show how generative language models can be combined with IoT in real-time. This allows for what was previously impossible: learning environments being able to not only feel conditions but also feel good – narrate them, and offer individualised explanations of why current atmospheric values, or experimental readings, matter for the particular learner in a particular context (Jayavardhini, 2024). Generative AI integrates IoT and makes sensors, not mere measurement devices, but Gen AI conversational agents.

6G and haptic internet: Current 5G-enabled smart classrooms have latencies ranging from 1 to 5 milliseconds – enough to deliver real-time video and adaptive assessment (Kar et al., 2021). Emerging 6G architectures promise sub-millisecond latency with deterministic quality-of-service guarantees, enabling haptic educational applications: remote laboratory manipulation with tactile feedback, collaborative physical construction across distance, and surgical and technical skill transmission requiring proprioceptive fidelity (Gupta et al., 2023).

Edge AI and privacy-preserving analytics: Emerging frameworks of educational AI focus on edge-deployed intelligence as an important privacy architecture. When AI processing takes place at the edge of the networks instead of at a centralised cloud, raw data video streams, biometric signals and interaction logs never need to leave the local environment. Model training and generating takes place on the device; only de-identified parameter generation propagates upwards (Gala, 2024). This is a deliberate architectural choice that turns privacy from being a matter of policy commitment to being an accident of technology.

3.9.2 Persistent Challenges and Ethical Considerations

Technological optimism must be tempered by recognition that barriers, structural in nature, can't just be dissolved by any kind of innovation.

Infrastructural inequality: The progression of reductions in latency and personalisation made possible by high connectivity smart classrooms is not available to a large segment of the world's schools, which are not equipped with reliable broadband connectivity. AI-IoT integration, without conscious efforts to integrate the systems equity-wise, will not reduce educational inequalities, but digitise them (Pillay et al., 2025). The classroom that can adapt to real-time changes of course material and the classroom that cannot adapt, and cannot even provide a functional internet connection, do not live in the same educational century.

Algorithmic bias amplification: Predictive methods, which are trained on historical educational data, tend to accept biases that are present in the data (Franklin et al., 2024). Systems that identify "at-risk" students are educated in a past in which some populations were systematically under-resourced and therefore under-performing. Put into use prospectively, such models may risk the transformation from descriptive historical inequity into prescriptive future destiny (Susnjak, 2023).

Data privacy and security: The multitude of sensors in the learning environments brings a very new attack surface. Attractive targets for malicious actors are biometric databases, continuous audiovisual databases and longitudinal behavioural records (Jimmy, 2024). Beyond the security questions, the conduct of the day-to-day functioning of these systems raises basic questions around the element of autonomy for students and the commodification of learning data.

Regulatory lag: Educational institutions are structured within the regulatory frameworks that are intended for similar situations. Many underlying privacy laws predate the internet, let alone ambient intelligence (Zainuddin, 2024). Subject to regulatory modernisation: Governance innovation has to be pursued in parallel with technical innovation; without modernisation of the regulations, smart classrooms will be in a perpetual state of legal ambiguity.

3.9.3 Redefining Educational Success

Perhaps the most transformative aspect of AI-IoT integration is that it serves to disrupt not only educational means but also educational ends. Standardised assessment, as the dominant measure of educational success since it was first introduced a hundred years ago, is a technology of scarcity. It came into existence as direct continuous observation of learning was not possible; indirect and periodic measurements were the best available alternative (Khlaif et al., 2024). Continuous assessment melts this scarcity, but has not yet melted the conceptual framework scarcity produced. What might educational success mean, without measurement in discrete time? This chapter, therefore, proposes three candidate dimensions.

Epistemic resilience: Not what students know when it is time to be measured, but how they bounce back from knowledge failure. Systems with the ability to track learning trajectories can measure not the peak performance but the recovery time, the time between the exposure to challenging material and the successful integration of that learning.

Collaborative capacity: Not an effluent achievement detached from the other, but a marginal contribution a learner makes in the construction of the collective knowledge. IoT AI systems that document group interaction allow group participation by a student to be evaluated with respect to increased group performance, not simply their individual score (Jony & Hamim, 2024).

Agency calibration: Not compliance with instructional directives, but sophistication in calibrating their help-seeking by learners. The student who knows when he/she needs assistance, how to ask for it so that they can get it effectively, and who integrates received help into independent performance shows a type of competency that is longer lasting than any amount of factual competence. These metrics are not operationalised yet – they are conceptual provocations. But they show the deeper opportunity of the integration of AI and IoT. The technology that frees education from the limitations of information scarcity may also free education from the limitations of its poverty-stricken definition of success.

3.10 CONCLUSION

This chapter has developed three interrelated propositions regarding the AI-IoT nexus and smart classrooms. First, the convergence of ambient sensing and AI is not building upon individual elements but qualitative transformation. The co-intelligent classroom is not different from its predecessors by degree but kind: the classroom knows things, infers things and adapts based on that information. It does not wait on instruction, but looks for need. Not an instrument, but a participant, would it be?

Second, this is a transformation in professional practice, which does not make educators obsolete. The role of the teacher changes from the role of knowledge transmission to that of pedagogical architecture – or designing the conditions in which intelligent systems can be used in support of learning. This shift requires new competencies yet presents new professional agency.

Third, there needs to be design, not retrofitting, of the ethical architecture of smart classrooms. Privacy as a concept cannot be added as a policy afterthought but needs to be inscribed in the logic of systems. Equity can't be remedied in supplemental programmes; it needs to be planned for in infrastructural investment. Human agency is something that cannot be taken for granted, instead it has to be consciously maintained within increasingly automated environments.

The term "smart classroom" has worn badly. Intelligence as currently operationalised in educational technology is still narrow – pattern recognition, but no understanding, adaptation without intention. The classrooms that the author writes about in this chapter are not yet intelligent in any human sense. Yet they are the signs of movement towards a condition which this chapter has called co-intelligence: the ability of human and artificial systems to be each doing what they are respectively most suitable for doing, in reciprocal interdependence. The environment sees what humans see, what human beings cannot see, in subtle physiological shifts, multivariate correlations in all of the thousands of learners. Humans interpret what the environment is not capable of interpreting – the meaning of a hesitant question, the meaning of a particular misconception for this very particular child. Neither is a substitute for the other, but one strengthens the other. This vision carries implications beyond classroom walls.

Educational institutions that are successful at growing co-intelligent practices may serve as a model for something that we feel society desperately needs: the integration of human and AI, not as competition but as collaboration. The smart classroom is not just any technical project. It is a prototype for civilisational co-existence with systems which think. These directors of the future are not just technologists. They are educators who are not on either side of the argument, but rather are educators who do not back either uncritical adoption or wholesale rejection – they are educators who insist that technology be in the service of pedagogical values rather than the determinant of them. They are policymakers who understand that infrastructure investment is equity investment. They are not just researchers of technical optimisation, but ethical investigation.

The classroom of the future will not be designed; it will be negotiated. Its contours will emerge from the collective choices which operate in and from it, building, funding, resisting and teaching. This chapter has tried to do no more than to bring out the stakes of such choices. The question before us is whether or not intelligent systems will make their way into classrooms. They have already arrived. The question is, though, what will we ask of them, what we will allow them to be and what we will preserve as inalienably human.

REFERENCES

Adewojo, A. A., & Dunmade, A. O. (2024). From big data to intelligent libraries: Leveraging analytics for enhanced user experiences. *Business Information Review*, *41*(3), 104–109. https://doi.org/10.1177/02663821241264707

Ajuwon, O., Animashaun, E., & Chiekezie, N. (2024). Integrating AI and technology in educational administration: Improving efficiency and educational quality. *Open Access Research Journal of Science and Technology*, *11*(2), 116–127. https://doi.org/10.53022/oarjst.2024.11.2.0102

Alexsius Pardosi, V. B., Xu, S., Umurohmi, U., Nurdiana, N., & Sabur, F. (2024). Implementation of an artificial intelligence-based learning management system for adaptive learning. Al-Fikrah: Jurnal Manajemen Pendidikan, *12*(1), 149. https://doi.org/10.31958/jaf.v12i1.12548

Al-Khresheh, M. H., & Alkursheh, T. O. (2024). An integrated model exploring the relationship between self-efficacy, technology integration via Blackboard, English proficiency, and Saudi EFL students' academic achievement. *Humanities and Social Sciences Communications*, *11*(1). https://doi.org/10.1057/s41599-024-02783-2

Al Mamun, M. A., & Lawrie, G. (2023). Student-content interactions: Exploring behavioural engagement with self-regulated inquiry-based online learning modules. *Smart Learning Environments*, *10*(1). https://doi.org/10.1186/s40561-022-00221-x

Alshanskaia, E. I., Zhozhikashvili, N. A., Polikanova, I. S., & Martynova, O. V. (2024). Heart rate response to cognitive load as a marker of depression and increased anxiety. *Frontiers in Psychiatry*, *15*. https://doi.org/10.3389/fpsyt.2024.1355846

Alzahrani, F. K., & Alhalafawy, W. S. (2023). Gamification for learning sustainability in the blackboard system: Motivators and obstacles from faculty Members' perspectives. *Sustainability*, *15*(5), 4613. https://doi.org/10.3390/su15054613

Amin, M. Y. M. (2023). AI and chat GPT in language teaching: Enhancing EFL classroom support and transforming assessment techniques. *International Journal of Higher Education Pedagogies*, *4*(4), 1–15. https://doi.org/10.33422/ijhep.v4i4.554

Antonijevic, M., Zivkovic, M., Djuric Jovicic, M., Nikolic, B., Perisic, J., Milovanovic, M., Jovanovic, L., Abdel-Salam, M., & Bacanin, N. (2025). Intrusion detection in the metaverse environment internet of things systems, by metaheuristics tuned two level framework. *Scientific Reports*, *15*(1). https://doi.org/10.1038/s41598-025-88135-9

Arifin, A., Suryaningsih, S., & Arifudin, O. (2024). The relationship between classroom environment, teacher professional development, and student academic performance in secondary education. *International Education Trend Issues*, *2*(2), 151–159. https://doi.org/10.56442/ieti.v2i2.467

Baena-Navarro, R., Carriazo-Regino, Y., Torres-Hoyos, F., & Pinedo-López, J. (2025). Intelligent prediction and continuous monitoring of water quality in aquaculture: Integration of machine learning and internet of things for sustainable management. *Water*, *17*(1), 82. https://doi.org/10.3390/w17010082

Boscardin, C. K., Abdulnour, R.-E. E., & Gin, B. C. (2025). Macy foundation innovation report Part I: Current landscape of artificial intelligence in medical education. *Academic Medicine: Journal of the Association of American Medical Colleges*, *100*(9S Suppl 1), S15–S21. https://doi.org/10.1097/acm.0000000000006107

Chauhan, B. (2024). AI-driven energy management systems for smart buildings. *Power System Technology*, *48*(1), 322–337. https://doi.org/10.52783/pst.280

Chiradeja, P., & Yoomak, S. (2023). Development of a public lighting system with smart lighting control systems and Internet of Things (IoT) technologies for a smart city. *Energy Reports*, *10*, 3355–3372. https://doi.org/10.1016/j.egyr.2023.10.027

Das, D., Bhattacharjee, N., & Boro, D. (2025). Real-time IoT data analysis for HVAC system maintenance. *Engineering Research Express*, *7*(3). https://doi.org/10.1088/2631-8695/ae0107

Dudek, S., Koziak, W., Makieła, M., Bętkowska, A., Kornacka, A., Dudek, W., Szostak, K., Tomaka, R., & Byra, A. (2025). Revolutionising sports: The role of wearable technology and AI in training and performance analysis. *Quality in Sport*, *39*, 58456. https://doi.org/10.12775/qs.2025.39.58456

Eden, C., Chisom, O., & Adeniyi, I. (2024). Integrating AI in education: Opportunities, challenges, and ethical considerations. *Magna Scientia Advanced Research and Reviews*, *10*(2), 006–013. https://doi.org/10.30574/msarr.2024.10.2.0039

Fazil, M., & Albahlal, B. M. (2025). MultiFAR: Multidimensional information fusion with attention-driven representation learning for student performance prediction. *PLoS One*, *20*(10). https://doi.org/10.1371/journal.pone.0333099

Fidalgo, P., & Thormann, J. (2024). *The future of lifelong learning: The role of artificial intelligence and distance education*. Intechopen. https://doi.org/10.5772/intechopen.114120

Franklin, G., Stephens, R., Piracha, M., Tiosano, S., Lehouillier, F., Koppel, R., & Elkin, P. L. (2024). The sociodemographic biases in machine learning algorithms: A biomedical informatics perspective. *Life (Basel, Switzerland)*, *14*(6), 652. https://doi.org/10.3390/life14060652

Gala, K. (2024). Ethical and legal considerations in AI-driven health cybersecurity. *International Journal of Scientific Research in Computer Science, Engineering and Information Technology*, *10*(5), 682–690. https://doi.org/10.32628/cseit241051059

Galesic, M., Barkoczi, D., Berdahl, A. M., Biro, D., Carbone, G., Giannoccaro, I., Goldstone, R. L., Gonzalez, C., Kandler, A., Kao, A. B., Kendal, R., Kline, M., Lee, E., Massari, G. F., Mesoudi, A., Olsson, H., Pescetelli, N., Sloman, S. J., Smaldino, P. E., & Stein, D. L. (2023). Beyond collective intelligence: Collective adaptation. *Journal of the Royal Society Interface*, *20*(200). https://doi.org/10.1098/rsif.2022.0736

Ghosh, L., & Ravichandran, R. (2024). Emerging technologies in vocational education and training. *Journal of Digital Learning and Education*, *4*(1), 41–49. https://doi.org/10.52562/jdle.v4i1.975

Gill, S. S., Xu, M., Patros, P., Wu, H., Kaur, R., Kaur, K., Fuller, S., Singh, M., Arora, P., Parlikad, A. K., Stankovski, V., Abraham, A., Ghosh, S. K., Lutfiyya, H., Kanhere, S. S., Bahsoon, R., Rana, O., Dustdar, S., Sakellariou, R., …, & Buyya, R. (2023). Transformative effects of ChatGPT on modern education: Emerging era of AI chatbots. *Internet of Things and Cyber-Physical Systems*, *4*, 19–23. https://doi.org/10.1016/j.iotcps.2023.06.002

Gkintoni, E., Antonopoulou, H., Sortwell, A., & Halkiopoulos, C. (2025). Challenging cognitive load theory: The role of educational neuroscience and artificial intelligence in redefining learning efficacy. *Brain Sciences*, *15*(2), 203. https://doi.org/10.3390/brainsci15020203

Gong, Y., Xu, C., Luo, S., & Lin, J. (2025). Modelling teacher education students' adoption of large language models through an extended technology acceptance framework. *Scientific Reports*, *15*(12). https://doi.org/10.1038/s41598-025-03298-9

Gordon, N., Al Jaber, T., & Omughelli, D. (2024). Fairness, bias, and ethics in AI: Exploring the factors affecting student performance. *Journal of Intelligent Communication*, *4*(1), 100–110. https://doi.org/10.54963/jic.v4i1.306

Gravett, K., & Lygo-Baker, S. (2024). Affective encounters in higher education. *Studies in Higher Education*, *50*(2), 228–238. https://doi.org/10.1080/03075079.2024.2332427

Gupta, M., Jha, R. K., & Jain, S. (2023). Tactile-Based Intelligence Touch Technology in IoT Configured WCN in B5G/6G-A Survey. *IEEE Access*, *11*, 30639–30689. https://doi.org/10.1109/access.2022.3148473

Imran, M., Almusharraf, N., Ahmed, S., & Mansoor, M. I. (2024). Personalisation of e-learning: Future trends, opportunities, and challenges. *International Journal of Interactive Mobile Technologies (iJIM)*, *18*(10), 4–18. https://doi.org/10.3991/ijim.v18i10.47053

Jetzinger, F., Baumer, S., & Michaeli, T. (2024). *Artificial Intelligence in compulsory K-12 computer science classrooms: A scalable professional development offer for computer science teachers. 37*, 590–596. https://doi.org/10.1145/3626252.3630782

Jimmy, F. (2024). Cyber security vulnerabilities and remediation through cloud security tools. *Journal of Artificial Intelligence General Science (JAIGS) ISSN:3006-4023*, *2*(1), 129–171. https://doi.org/10.60087/jaigs.v2i1.102

Jony, A. I., & Hamim, S. A. (2024). Empowering virtual collaboration: Harnessing AI for enhanced teamwork in higher education. *Educational Technology Quarterly*, *2024*(3), 337–359. https://doi.org/10.55056/etq.746

Kamil, Z., & Abdulazeez, A. M. (2024). A review on deep reinforcement learning for autonomous driving. *Indonesian Journal of Computer Science*, *13*(3). https://doi.org/10.33022/ijcs.v13i3.4036

Kar, S., Mishra, P., & Wang, K.-C. (2021, October 1). *5G-IoT architecture for next generation smart systems*. https://doi.org/10.1109/5gwf52925.2021.00049

Khlaif, Z., Odeh, A., & Bsharat, T. R. K. (2024). *Generative AI-powered adaptive assessment* (pp. 157–176). IGI Global. https://doi.org/10.4018/979-8-3693-6397-3.ch007

Kim, J., Detrick, R., Yu, S., Song, Y., Bol, L., & Li, N. (2025). Socially shared regulation of learning and artificial intelligence: Opportunities to support socially shared regulation. *Education and Information Technologies*, *30*(9), 11483–11521. https://doi.org/10.1007/s10639-024-13187-9

Klęczek, K., Rice, A., & Alimardani, M. (2024). Robots as mental health coaches: A study of emotional responses to technology-assisted stress management tasks using physiological signals. *Sensors (Basel, Switzerland)*, *24*(13), 4032. https://doi.org/10.3390/s24134032

Knights, V., Petrovska, O., & Prchkovska, M. (2024). *Enhancing smart parking management through machine learning and AI integration in IoT environments.* Intechopen. https://doi.org/10.5772/intechopen.1006490

Koukaras, C., Koukaras, P., Ioannidis, D., & Stavrinides, S. G. (2025). AI-driven telecommunications for smart classrooms: Transforming education through personalised learning and secure networks. *Telecom*, *6*(2), 21. https://doi.org/10.3390/telecom6020021

Krumsvik, R. J. (2025). GPT-4's capabilities for formative and summative assessments in Norwegian medicine exams-an intrinsic case study in the early phase of intervention. *Frontiers in Medicine*, *12*. https://doi.org/10.3389/fmed.2025.1441747

Kwarteng, S., Appiagyei, E., Larbi, E., & Lotey, E. K. (2025). Predicting mathematics Facilitators' teaching efficacy through their engagement in professional learning communities: A mixed-method approach. *Journal of Mathematics, Science and Technology Education*, *2*(2), 117–128. https://doi.org/10.12973/jmste.2.2.117

Lakhe Shrestha, B. L., Dahal, N., Hasan, M. K., Paudel, S., & Kapar, H. (2025). Generative AI on professional development: A narrative inquiry using the TPACK framework. *Frontiers in Education*, *10*. https://doi.org/10.3389/feduc.2025.1550773

Li, D. (2024). Creating a personalised higher education teaching system using fuzzy association rule mining. *International Journal of Computational Intelligence Systems*, *17*(1). https://doi.org/10.1007/s44196-024-00641-2

Long, C., Sam, R., Ny, C., Chhang, C., Ren, R., Ngork, C., Sorn, R., Sorn, M., & Sor, C. (2024). The impact of assessment for 21st century skills in higher education Institutions: A narrative literature review. *International Journal of Advanced Social Sciences and Education (IJASSE)*, *2*(1), 19–42. https://doi.org/10.59890/ijasse.v2i1.1378

Lunn Brownlee, J., Bourke, T., Rowan, L., Ryan, M., Churchward, P., Walker, S., L'Estrange, L., Berge, A., & Johansson, E. (2022). How epistemic reflexivity enables teacher educators' teaching *for* diversity: Exploring a pedagogical framework for critical thinking. *British Educational Research Journal*, *48*(4), 684–703. https://doi.org/10.1002/berj.3789

Luo, Q. Z. (2023). The influence of AI-powered adaptive learning platforms on student performance in Chinese classrooms. *Journal of Education*, *6*(3), 1–12. https://doi.org/10.53819/81018102t4181

Mahmood, N., Bhatti, S. M., Dawood, H., Pradhan, M. R., & Ahmad, H. (2024). Measuring student engagement through behavioural and emotional features using deep-learning models. *Algorithms*, *17*(10), 458. https://doi.org/10.3390/a17100458

Maity, S., & Deroy, A. (2024). *Generative AI and its impact on personalised intelligent tutoring systems.* Centre for Open Science. https://doi.org/10.35542/osf.io/kawr5

Mehta, P., Chillarge, G. R., Sapkal, S. D., Shinde, G. R., & Kshirsagar, P. S. (2023). *Inclusion of children with special needs in the educational system, Artificial Intelligence (AI)* (pp. 156–185). IGI Global. https://doi.org/10.4018/979-8-3693-0378-8.ch007

Mishra, R., & Varshney, D. (2024). Comprehensive analysis of human and AI task allocation in the education sector: Defining futuristic roles and responsibilities. *World Journal of Advanced Research and Reviews*, *22*(3), 1883–1893. https://doi.org/10.30574/wjarr.2024.22.3.1949

Mousavi, D., Ghaemian, M., & Brinkerhoff, J. (2024). Multiscale CFD analysis of urban air pollution dome and ventilation enhancement via an urban chimney. *Atmospheric Environment*, *337*, 120783. https://doi.org/10.1016/j.atmosenv.2024.120783

Naciri, A., Radid, M., Kharbach, A., & Chemsi, G. (2021). E-learning in health professions education during the COVID-19 pandemic: A systematic review. *Journal of Educational Evaluation for Health Professions*, *18*, 27. https://doi.org/10.3352/jeehp.2021.18.27

Naseer, F., Khan, M. N., Addas, A., Awais, Q., & Ayub, N. (2025). Game mechanics and artificial intelligence personalisation: A framework for adaptive learning systems. *Education Sciences*, *15*(3), 301. https://doi.org/10.3390/educsci15030301

Naseer, F., & Khawaja, S. (2025). Mitigating conceptual learning gaps in mixed-ability classrooms: A learning analytics-based evaluation of AI-driven adaptive feedback for struggling learners. *Applied Sciences*, *15*(8), 4473. https://doi.org/10.3390/app15084473

Osman, S. A., & Ahmed, Z. E. (2024). *Navigating AI integration* (pp. 240–267). IGI Global. https://doi.org/10.4018/979-8-3693-2728-9.ch011

Jayavardhini, P. (2024). AI study partner : Development of an LLM and gen AI-enhanced study assistant tool. *International Journal of Scientific Research in Engineering and Management*, *08*(03), 1–5. https://doi.org/10.55041/ijsrem29626

Pillay, T. S., Khan, A. I., & Yenice, S. (2025). Artificial intelligence (AI) in point-of-care testing. *Clinica Chimica Acta; International Journal of Clinical Chemistry*, *574*, 120341. https://doi.org/10.1016/j.cca.2025.120341

Ragab, A., Hassieb, M. M., & Mohamed, A. F. (2025). Exploring the impact of window design and ventilation strategies on air quality and thermal comfort in arid educational buildings. *Scientific Reports*, *15*(1). https://doi.org/10.1038/s41598-025-01362-y

Rajput, R. (2025). *Use of artificial intelligence to solve problems in the classroom* (pp. 137–168). IGI Global. https://doi.org/10.4018/979-8-3373-0437-3.ch005

Ramakrishnan, S., Bishnoi, M. M., Joghee, S., Jijitha, S., & Kumar, A. (2024). *Social engineering: Role of teachers in cohabitation of AI with education*. 1–6. https://doi.org/10.1109/iccr61006.2024.10532897

Rathika, P., Yamunadevi, S., Ponni, P., Parthipan, V., & Anju, P. (2024). Developing an AI-powered interactive virtual tutor for enhanced learning experiences. *International Journal of Computational and Experimental Science and Engineering*, *10*(4). https://doi.org/10.22399/ijcesen.782

Rosenbaum, S., Moberg, J., Chesire, F., Mugisha, M., Ssenyonga, R., Ochieng, M. A., Simbi, C. M. C., Nakyejwe, E., Ngatia, B., Rada, G., Vásquez-Laval, J., Garrido, J. D., Baguma, G., Kuloba, S., Sebukyu, E., Kabanda, R., Mwenyango, I., Muzaale, T., Nandi, P., … , & Oxman, M. (2024). Teaching critical thinking about health information and choices in secondary schools: Human-centred design of digital resources. *F1000Research*, *12*, 481. https://doi.org/10.12688/f1000research.132580.2

Sajadieh, S. M. M., & Noh, S. D. (2025). From simulation to autonomy: Reviews of the integration of artificial intelligence and digital twins. *International Journal of Precision Engineering and Manufacturing-Green Technology*, *12*(5), 1597–1628. https://doi.org/10.1007/s40684-025-00750-z

Saputra, I., Kurniawan, A., Yanita, M., Yeni Putri, E., & Mahniza, M. (2024). The evolution of educational assessment: How artificial intelligence is shaping the trends and future of learning evaluation. *The Indonesian Journal of Computer Science*, *13*(6). https://doi.org/10.33022/ijcs.v13i6.4465

Sasal, L., Chakraborty, T., & Hadid, A. (2022). W-transformers: A wavelet-based transformer framework for univariate time series forecasting. *2022 21st IEEE International Conference on Machine Learning and Applications (ICMLA)*. *34*, 671–676. https://doi.org/10.1109/icmla55696.2022.00111

Seraj, M., Parvez, M., Khan, O., & Yahya, Z. (2024). Optimising smart building energy management systems through Industry 4.0: A response surface methodology approach. *Green Technologies and Sustainability*, *2*(2), 100079. https://doi.org/10.1016/j.grets.2024.100079

Sidorkin, A. M. (2025). AI Integration Blueprint. *AI-EDU Arxiv*. https://doi.org/10.36851/ai-edu.vi.5126

Sieg, L., Eismann, H., Schneider, B., Karsten, J., & Darici, D. (2025). How distractions undermine attentional synchronisation in work-based learning: A randomised controlled study using dual Mobile eye-tracking. *Journal of Computer Assisted Learning*, *41*(3). https://doi.org/10.1111/jcal.70059

Sudan, T., Hans, A., & Taggar, R. (2024). Transformative learning with ChatGPT: Analysing adoption trends and implications for business management students in India. *Interactive Technology and Smart Education*, *21*(4), 735–772. https://doi.org/10.1108/itse-10-2023-0202

Susnjak, T. (2023). Beyond predictive learning analytics modelling and onto explainable artificial intelligence with prescriptive analytics and ChatGPT. *International Journal of Artificial Intelligence in Education*, *34*(2), 452–482. https://doi.org/10.1007/s40593-023-00336-3

Thuy, P. B., & Tien, P. D. (2024). AI-powered administration: The role of intelligent tutoring systems in education. *International Journal of Religion*, *5*(10), 4560–4569. https://doi.org/10.61707/sqwrjn32

Tutul, A. A., Nirjhar, E. H., & Chaspari, T. (2024). Investigating trust in human-AI collaboration for a speech-based data analytics task. *International Journal of Human–Computer Interaction*, *41*(5), 1–19. https://doi.org/10.1080/10447318.2024.2328910

Viberg, O., Cukurova, M., Feldman-Maggor, Y., Alexandron, G., Shirai, S., Kanemune, S., Wasson, B., Tømte, C., Spikol, D., Milrad, M., Coelho, R., & Kizilcec, R. F. (2024). What explains Teachers' trust in AI in education across six countries? *International Journal of Artificial Intelligence in Education*, *35*(3), 1288–1316. https://doi.org/10.1007/s40593-024-00433-x

Vulic, J., Jacobson, M. J., & Levin, J. A. (2024). Exploring education as a complex system: Computational educational research with multi-level agent-based modelling. *Education Sciences*, *14*(5), 551. https://doi.org/10.3390/educsci14050551

Wang, Z., Yan, W., Zeng, C., Tian, Y., & Dong, S. (2023). A unified interpretable intelligent learning diagnosis framework for learning performance prediction in intelligent tutoring systems. *International Journal of Intelligent Systems*, *2023*(1), 1–20. https://doi.org/10.1155/2023/4468025

Wen, J., Nie, J., Zhong, Y., Yi, C., Li, X., Jin, J., Zhang, Y., & Niyato, D. (2024). Diffusion-model-based incentive mechanism with Prospect theory for edge AIGC services in 6G IoT. *IEEE Internet of Things Journal*, *11*(21), 34187–34201. https://doi.org/10.1109/jiot.2024.3445171

Wu, Y., & Yusof, Y. (2024). Emerging trends in real-time recommendation systems: A deep dive into multi-behaviour streaming processing and recommendation for E-commerce platforms. *Journal of Internet Services and Information Security*, *14*(4), 45–66. https://doi.org/10.58346/jisis.2024.i4.003

Xiao, N., Pei, Y., Yuan, C., Bu, Y., & Cai, Z. (2025). Transforming education with artificial intelligence: A comprehensive review of applications, challenges, and future directions. *International Theory and Practice in Humanities and Social Sciences*, *2*(1), 337–356. https://doi.org/10.70693/itphss.v2i1.211

Yan, D. (2024). Feedback-seeking abilities of L2 writers using ChatGPT: A mixed-method multiple case study. *Kybernetes*, *54*(7), 3757–3781. https://doi.org/10.1108/k-09-2023-1933

Zainuddin, N. (2024). Does artificial intelligence cause more harm than good in schools? *International Journal of Language Education and Applied Linguistics*, *14*(1), 1–3. https://doi.org/10.15282/ijleal.v14i1.10432

Zhao, Y., & Wang, L. (2022). A case study of student development across project-based learning units in middle school chemistry. *Disciplinary and Interdisciplinary Science Education Research*, *4*(1). https://doi.org/10.1186/s43031-021-00045-8

Zou, Y., Kuek, F., Feng, W., & Cheng, X. (2025). Digital learning in the 21st century: Trends, challenges, and innovations in technology integration. *Frontiers in Education*, *10*. https://doi.org/10.3389/feduc.2025.1562391

4 Prompt Literacy for Educators

A Framework for Effective AI Interaction in Lesson Planning

Adil El jouali and Wiame El amri

4.1 INTRODUCTION

In recent years, AI has completely changed several aspects of our daily lives. In the education field, Generative Artificial Intelligence (GenAI) is becoming indispensable to educators. However, many of them still don't fully understand the basics of prompt construction, which impacts the quality of outputs.

This link between the quality of prompts and the quality of outputs is especially important for lesson planning, which requires upfront specification of key pedagogical parameters (objectives, student profiles, assessment, differentiation, scope).

This study addresses the limits of existing prompt engineering guides through a three-phased approach. First, we evaluate six major guides (ChatGPT, Gemini, Claude, Harvard University, University of Michigan, University of Utah, MagicSchool). Second, we assess five real life prompts collected from publicly accessible online platforms (e.g., Reddit). Finally, we examine the corresponding ChatGPT-4 generated outputs. Our findings show that there is a disconnect between pedagogical logic and existing guidance, offering limited help to educators. Based on this, we propose the CRAFT framework, a pedagogically grounded approach for AI-assisted lesson planning.

4.2 PROMPT LITERACY IN EDUCATIONAL PRACTICE: A LITERATURE REVIEW

4.2.1 Prompt Literacy: Concept and Frameworks

Prompt literacy is a cyclical process that involves the ability to effectively formulate, interpret and iteratively refine prompts to get appropriate AI responses (Gattupalli et al., 2023; Hwang et al., 2023; Tour and Zadorozhnyy, 2025).

For educators, prompt literacy means being able to use AI to achieve specific educational outcomes. As such, it requires pedagogical content knowledge and awareness of student diversity (Warschauer et al., 2023).

There are many frameworks for prompt engineering, including Lo's (2023) CLEAR model, and the framework proposed by Korzynski et al.'s (2023) which highlights context, instructions, input data and output format. However, these frameworks describe a good prompt instead of guiding users on building one. To address this gap, Tour and Zadorozhnyy (2025) draw on Freebody and Luke's (1999) Four Resources Model and adapt it to AI prompting. Their adapted model focuses on understanding basic prompting practices, creating effective prompts and critically analyzing both prompts and AI outputs. By moving beyond technical refinement, it promotes a more reflective form of prompt literacy that keeps AI assisted lesson planning pedagogically sound.

4.2.2 GenAI and Lesson Planning: Opportunities and Challenges

Recent advances in GenAI have opened up new opportunities for educators. These range from improving lesson design and assessment (Octavio et al., 2024) to reducing planning time and

DOI: 10.1201/9781003767855-4

mitigating burnout (Hashem et al., 2024) and creating interactive and differentiated materials (Yeh, 2024). However, the value of AI depends entirely on the quality of prompts (Husomanovic, 2025; Kehoe, 2023).

Research generally agrees on what makes an effective prompt for lesson planning. Key elements include clear learning goals (Lee and Palmer, 2025; Yeh, 2024), clear student descriptions (Octavio et al., 2024), differentiation (Yeh, 2024), and realistic scope and time constraints (Kehoe, 2023). However, even when all these elements are present, AI outputs still face some problems, such as fake references, repetitive content and limited context awareness (Kehoe, 2023; van den Berg and du Plessis, 2023). These challenges reinforce the idea that AI-generated lesson plans should be seen as preliminary drafts, requiring careful review and adaptation by educators.

4.2.3 Methodology

This study follows a qualitative three-phased approach. First, we compared prompt engineering guidance from six major and diverse sources, from technology companies (Open AI, Google Gemini, Anthropic Claude), to major universities (Harvard, Michigan, Utah) and education platforms (MagicSchool.ai). Each one was evaluated across six dimensions (target audience, lesson planning grounding, comprehensiveness, accessibility, actionability and clarity, skill development). Second, we collected five real life lesson planning prompts from publicly accessible spaces (Reddit, education forums, Facebook groups). Selection was based on explicit focus on lesson planning, completeness and diverse coverage (grade levels, subject areas, institutional contexts). Each prompt was then evaluated on a 0–4 grid based on instructional design principles identified as essential for lesson planning in the literature, including pedagogical specificity (grade level, curriculum context, student characteristics), instructional design elements (objectives, assessment, differentiation, prior knowledge, scaffolding) and prompt construction quality (role definition, task clarity, format specification, constraints). Finally, original prompts were entered into fresh ChatGPT-4 sessions to generate baselines outputs. Based on gaps identified, prompts were then systematically revised, enriched and entered into new ChatGPT-4 sessions to prevent carryover effects. Finally, we compared and scored outputs across five dimensions (instructional soundness, standards alignment, differentiation adequacy, assessment validity, context appropriateness).

4.2.3.1 Comparative Analysis of AI Prompting Guides

Over the last few years, major AI developers and educational institutions have developed increasingly accessible guidance for prompt engineering. This analysis synthesizes guidance from six major sources representing diverse stakeholders: Open AI's ChatGPT, Harvard University, University of Michigan, University of Utah, Google's Gemini, Anthropic's Claude and MagicSchool.ai (Table 4.1).

By analyzing the different technical and institutional prompt engineering guides, we find that there is an important misalignment between them and educator needs across three dimensions:

1. Pedagogical grounding: these guidelines articulate universal principles but lack pedagogical reasoning. Emphasis on specificity is more confusing than it is helpful as it leaves educators with the task of translating generic advice without specifying which pedagogical details matter. Moreover, none of the guides mention pre-prompt pedagogical thinking (objectives, student diversity, standards, assessments).
2. Operational clarity: technical guidelines like those provided by OpenAI and Gemini (which are coincidently the most commonly used ones by educators and non-educators alike) are mainly targeted at audiences with technical expertise, offering limited guidance for normal people, educators included. MagicSchool represents the opposite extreme, with perfect operational clarity for educators through pre-structured tools. However, in doing so, the platform builds platform dependency as educators develop tool proficiency without understanding why certain details matter.

TABLE 4.1
Comparative Analysis of Prompt Engineering Guides

Dimension	OpenAI	Harvard	U-Michigan	U-Utah	Google Gemini	Anthropic Claude	MagicSchool
Target audience	General users, education as one use case	Academic users, research-focused	Campus community, practical but not pedagogy-specific	Educators at all levels, moderate pedagogical grounding	Developers, technical focus	Developers	K-12 educators, strong pedagogical focus
Lesson planning grounding	No reference to learning objectives, standards, differentiation	No scaffolding for instructional design	No teaching/learning theory reference	Emphasizes domain knowledge but lacks explicit lesson planning scaffolding	Misses pedagogical dimensions entirely	No educational guidance	Explicit lesson planning tools, fields for standards, differentiation, assessment
Comprehensiveness	Partial: covers construction but ignores pre-prompt pedagogical reasoning	Partial: strong on techniques but weak on pedagogical entry	Partial: missing structured approach, evaluation framework	Moderate-High: addresses domain knowledge and ethics but not operationalized	Low: technical focus, irrelevant to lesson planning	Low: optimized for coding workflows	High: operationalized for lesson planning, covers construction and evaluation
Educator accessibility	Moderate: requires translation, no K-12 specificity	Low-Moderate: academic examples don't translate	Moderate: generalness obscures classroom realities	Moderate-High: considers constraints but principles remain abstract	Low: developer audience, technical jargon	Low: technical complexity	High: K-12 language, teacher workflows, +80 tools
Clarity and actionability	Conversational but unstructured, high cognitive load	Organized but dense, medium-high load	Scattered tips, high load	Numbered progression, medium load	Very high load, overwhelming complexity	Very high load, intimidating	Very low load, form-filling with scaffolds, immediate implementation
Skill development	Builds transferable principles but moderate literacy development	Good literacy development but no systematic pathway	Good literacy but principles are scattered	Good literacy with domain knowledge emphasis	Advanced technical literacy for developers	Advanced technical literacy	Minimal literacy development, tool proficiency without understanding, platform-dependent

3. Transferable literacy: none of the guides enable educators to function as AI literate instructional designers across platforms while preserving sound pedagogy. General guidelines build cross-platform skills but lack pedagogical value, while educational platforms limit transferability.

This highlights the lack of a unified, pedagogically grounded prompt literacy framework, one that:

- Supports pedagogical reasoning
- Offers a clear and memorable structure
- Works across AI platforms
- Builds understanding rather than simple tool proficiency
- Places educator judgment at the center

4.2.4 Evaluation of Prompts and Outputs

4.2.4.1 Prompt Assessment

After selecting five real life lesson planning prompts from publicly accessible online spaces (Reddit, Facebook groups, educator forums), we evaluated each of them on a scale from 0 to 4 across three dimensions: pedagogical specificity, instructional design elements, prompt construction quality (Table 4.2).

TABLE 4.2
Original Prompt Evaluation

Prompt ID	Pedagogical Specificity (0–4)	Instructional Design Elements (0–4)	Prompt Construction Quality (0–4)	Total Score (0–12)	Critical Gaps
A	3/4 (Grade *Y*, Curriculum *Y*, Standards *N*, Student context *N*)	1/4 (Assessment *N*, Differentiation *N*, Prior knowledge *V*, Scaffolding *V*)	3/4 (Role *V*, Task *Y*, Format *Y*, Constraints *Y*)	7/12	No learning objectives, no student characteristics, content-coverage focus not learning-outcomes
B	2/4 (Grade *Y*, Curriculum *Y*, Standards *V*, Subject/topic *N*)	2/4 (Assessment *N*, Differentiation *N*, Prior knowledge *N*, Scaffolding *Y*)	3/4 (Role *Y*, Task *V*, Format *Y*, Constraints *Y*)	7/12	No subject/topic specified, no student characteristics, teaching style defined but student needs absent
C	2/4 (Audience *V*, Duration *Y*, Standards *N*, Student context *N*)	3/4 (Assessment *Y*, Differentiation *V*, Prior knowledge *V*, Scaffolding *Y*)	4/4 (Role *V*, Task *Y*, Format *Y*, Constraints *Y*)	9/12	Target audience too vague, assumes homogeneous learners, one-size-fits-all design despite technical sophistication
D	2/4 (Grade *Y*, Objective *Y*, Curriculum *N*, Student context *N*)	0/4 (Assessment *N*, Differentiation *N*, Prior knowledge *N*, Scaffolding *N*)	1/4 (Role *N*, Task *Y*, Format *N*, Constraints *N*)	3/12	Minimal prompt, maximum AI interpretation, no prior knowledge, student characteristics, time frame, scope, assessment, differentiation
E	1/4 (Course *V*, Grade *N*, Standards *N*, Student context *V*)	3/4 (Assessment *Y*, Differentiation *N*, Prior knowledge *Y*, Scaffolding *Y*)	4/4 (Role *Y*, Task *Y*, Format *Y*, Constraints *Y*)	8/12	Assumes homogeneity, no accommodation for learning differences, disabilities, language proficiency

Analyzing these prompts reveals systematic pedagogical gaps that persist regardless of how sophisticated the prompt is. All five prompts specify grade level, demonstrating educators understand basic context requirements, yet none mention student characteristics beyond broad demographic categories. Student ability diversity, prior knowledge and specific learning needs remain entirely absent. Assessment strategy appears in only one prompt and even then remains abstract rather than operationalized. Differentiation is completely absent across all prompts. Most critically, prompts focus overwhelmingly on content coverage (listing topics, chapters, or grammar points) rather than articulating learning outcomes as measurable student capabilities.

These patterns may reflect several factors beyond lack of framework. Educators may deliberately use minimal prompts as exploratory strategies to see what AI generates without constraints, or may face time pressures that discourage lengthy prompt construction. Privacy concerns may prevent sharing personal student details in widely accessible AI systems. Some educators may assume AI systems possess contextual knowledge about educational settings that would make explicit specification unnecessary. Critically, these patterns demonstrate that existing guidance successfully teaches prompt construction mechanics but does not effectively operationalize what pedagogical details matter for instructional design, nor does it help educators recognize that AI systems require explicit articulation of information humans with shared professional knowledge consider contextually obvious.

4.2.4.2 Output Analysis

For this phase of our analysis, the original prompts were entered into fresh ChatGPT-4 sessions to generate baseline outputs. Prompts were then revised and enhanced with pedagogical detail and reentered into new sessions. Outputs (original and improved) were compared and scored across instructional soundness, standards alignment, differentiation adequacy, assessment validity, context appropriateness (Table 4.3).

- Prompt A (5th grade science, musculoskeletal system): the original output listed eleven topics for ten lessons without measurable objectives, focused on content coverage rather than learning outcomes, referenced curriculum superficially without demonstrated alignment, included no differentiation despite mentioning mixed abilities and provided no assessment strategy beyond mentioning a future project. On the other hand, the improved output progressed logically, expressed objectives as measurable capabilities, operationalized differentiation with concrete strategies (i.e., visual supports) and included clear criteria-based assessment throughout.

TABLE 4.3
Comparative Output Analysis

Prompt ID	Version	Instructional Soundness	Standards Alignment	Differentiation Adequacy	Assessment Validity	Context Appropriateness
A	Original	2	3	1	1	1
	Improved	5	5	5	5	5
B	Original[a]	5	Cannot assess	Cannot assess	Cannot assess	5
	Improved	5	5	5	5	5
C	Original	4	2	2	4	2
	Improved	4	3	4	3	5
D	Original	3	1	3	2	2
	Improved	5	4	5	5	5
E	Original	4	2	1	4	2
	Improved	5	3	4	5	5

[a] For prompt B Original, ChatGPT-4 refused generation and requested essential information.

- Prompt B (5th grade science, states of matter): the original prompt, despite sophisticated specification of pedagogical model (Churer) and curriculum framework (LehrplanPLUS Bayern), received a response from ChatGPT-4 that appropriately refuses generation and requested five essential pieces: subject, topic, textbook, duration and special conditions. The improved output, once these gaps were filled, provided a complete 90 min lesson structured via the Churner model with explicit time allocations, promoted scientific thinking through guiding questions, aligned explicitly to LehrplanPLUS competencies, operationalized differentiation for mixed abilities including specific supports for a German language learner and included formative checkpoints with clear purposes.
- Prompt C (community college economics): the original output produced a well-structured nine week course with logical progression and multiple assessment types, but assumed homogenous students without accommodation for working students, math anxiety, disabilities or language proficiency needs. The output was a one-size-fits-all design inappropriate for community college contexts. The improved output acknowledged student realities through plain language and real world emphasis in module titles, respected math anxiety by starting with accessible concepts, employed collaborative approach and explicitly designed for students working part-time. However, this prompt received only modest improvements in standards alignment because the self-designed curriculum was not tied to specific accreditation standards.
- Prompt D (year 8 German, perfect tense): the original output showed reasonable lesson structure (started, presentation, guided practice, independent practice, plenary) but unrealistic scope attempting to introduce concept, practice formation and expect production within 60 min. differentiation was superficial rather than addressing diverse learning needs specifically. Assessment lacked clarity on what mastery looks like. The improved output established realistic 50 min scope focusing only on recognition and formation of five regular verbs, provided explicit time allocations, set appropriate production expectations, operationalized differentiation for struggling students/confident students/dyslexia needs and designed assessment with explicit purpose.
- Prompt E (self-directed economics): the original output showed excellent four year logical progression with realistic workload. However, it assumed learner homogeneity with no accommodation for special needs and learning differences, and assessment relied on so-called "AI professor." The improved output acknowledge specific learner needs (i.e., math anxiety), provided self-evaluation benchmarks, offered flexible pacing and suggested peer support communities. Assessment included quizzes with detailed explanations of mastery indicators.

There's a strong link between pedagogical specificity and the quality of output. Student characteristics and differentiation are critical for context appropriate AI-generated lesson plans. Measurable learning goals are equally important, as well as assessment, which has to be explicitly specified to ensure evaluation mechanisms that are coherent, aligned to objectives and with clear success criteria. Finally, scope and timing constraints are essential to frame output and ensure it remains realistic and feasible.

4.2.5 The CRAFT Framework

Since existing frameworks don't provide explicit and structured guidance on how to construct a good prompt with embedded pedagogical reasoning (Tour and Zadorozhnyy, 2025), we propose the CRAFT framework: Contextualizing learners, Refining objectives, Aligning assessment, Framing differentiation, Timing instruction.

Contextualizing learners specifies student characteristics (e.g., class size, ability levels, prior knowledge and specific needs). Refining objectives outlines measurable learning goals. Aligning assessment defines success criteria and connects them to learning objectives and future instruction. Framing differentiation extends contextualizing learners by outlining scaffolding with concrete materials, identifying potential extensions and establishing multiple entry points. Timing instruction entails setting realistic scope and timing constraints by specifying content coverage, defining timeframes, institutional constraints and accounting for available resources.

The framework operationalizes these elements through a three-stage process:

1. Pre-prompt pedagogical planning: this phase addresses class composition, ability ranges, prior knowledge, specific needs, measurable learning objectives, assessment methods and constraints (time, scope, resources, institutional).
2. Prompt construction: this phase incorporates each of the framework's components (student population description, measurable learning objectives, constraints, assessment specifications, differentiation requirements and output guidance.
3. Output evaluation: this phase involves assessing AI-generate output against explicit criteria (alignment with objectives, operationalization of differentiation, realistic time allocations, etc.).

By frontloading specification in stages 1 and 2, enabling educators to receive higher quality initial outputs that require slight adjustments rather than whole revision. We position our framework as one systematic approach among legitimate alternatives, educators preferring emergent, iterative processes may find value in prompting strategies emphasizing exploration and co-construction with AI over multiple exchanges. CRAFT is pedagogically complete, designed to be platform-agnostic and positions AI as a design partner rather than an implementation tool.

4.3 CONCLUSION

As GenAI increasingly infiltrates educational practice, prompt literacy imposes itself as an indispensable competency. Existing guidance on AI interaction fail to connect prompt engineering principles to pedagogical relevance, a pattern consistent across all six major guides covered in our analysis. Pedagogical specificity strongly influences output quality. Concrete student descriptions, measurable objectives, explicit assessment purposes, operationalized differentiation and realistic scope boundaries consistently produced more instructionally sound, contextually appropriate outputs than technically sophisticated prompts lacking these elements. We propose the CRAFT framework, a structured approach that directly embeds lesson planning logic into prompt construction through five components. Contextualizing learners specifies student characteristics, such as class size, ability levels, prior knowledge and specific needs. Refining objectives defines measurable learning goals. Aligning assessment specifies success criteria and connects it to objectives and future instruction. Framing differentiation extends contextualizing learners by outlining scaffolding with concrete materials, identifying potential extensions and establishing multiple entry points. Timing instruction entails setting realistic scope and timing constraints by specifying content coverage, defining timeframes, institutional constraints and accounting for available resources.

The model reflects the workflow of instructional design through a three-step operationalization process (pre-prompt pedagogical planning, prompt construction, output evaluation), therefore positioning AI as a co-planning partner while preserving human judgment and building transferable literacy.

However, our study faces several limitations. First, the generalizability of our findings is restricted by the small sample of prompts and the single use of ChatGPT-4. Second, our analysis relied on single attempt outputs rather than iterative refinement processes typically used by AI users. Finally, scoring and quality assessments reflect our pedagogical expertise and involve inherent subjectivity. Future research should test CRAFT across multiple AI platforms.

REFERENCES

Freebody, P., & Luke, A. *Further notes on the four resources model.* Reading Online, International Reading Association, 1999. http://www.readingonline.org

Gattupalli, S., Maloy, R. W., & Edwards, S. A. *Prompt literacy: A pivotal educational skill in the age of AI.* College of Education Working Papers and Reports Series, 6, 2023. https://doi.org/10.7275/3498-wx48

Hashem, R., Ali, N., El Zein, F., Fidalgo, P., & Abu Khurma, O. *AI to the rescue: Exploring the potential of ChatGPT as a teacher ally for workload relief and burnout prevention.* Research and Practice in Technology Enhanced Learning, 19, 023, 2024. https://doi.org/10.58459/rptel.2024.19023

Husomanovic, D. *Prompt literacy: The new basic skill for adults?* EPALE, 2025. https://epale.ec.europa.eu/en/blog/prompt-literacy-new-basic-skill-adults

Hwang, Y., Lee, J. H., & Shin, D. *What is prompt literacy? An exploratory study of language learners' development of new literacy skill using generative AI.* arXiv, 2023. https://doi.org/10.48550/arXiv.2311.05373

Kehoe, F. *Leveraging generative AI tools for enhanced lesson planning in initial teacher education at post-primary.* Irish Journal of Technology Enhanced Learning, 2023. https://doi.org/10.22554/ijtel.v7i2.124

Korzynski, P., Mazurek, G., Krzypkowska, P., & Kurasinski, A. *Artificial intelligence prompt engineering as a new digital competence: Analysis of generative AI technologies such as ChatGPT.* Entrepreneurial Business and Economics Review, 11(3), 25–37, 2023. https://www.ceeol.com/search/article-detail?id=1205908

Lee, D., & Palmer, E. *Prompt engineering in higher education: A systematic review to help inform curricula.* International Journal of Educational Technology in Higher Education, 22, Article 7, 2025. https://doi.org/10.1186/s41239-025-00503-7

Lo, L. S. *The CLEAR path: A framework for enhancing information literacy through prompt engineering.* Journal of Academic Librarianship, 49(4), 102720, 2023. https://www.sciencedirect.com/science/article/pii/S0099133323000599

Octavio, M. M., Argüello, M. V. G., & Pujolà, J.-T. *ChatGPT as an AI L2 teaching support: A case study of an EFL teacher.* Technology in Language Teaching & Learning, 6(1), 1–25, 2024. https://doi.org/10.29140/tltl.v6n1.1142

Tour, E., & Zadorozhnyy, A. *Conceptualizing and operationalizing prompt literacy for English language learners.* Journal of Adolescent & Adult Literacy, 2025. https://doi.org/10.1002/jaal.70020

van den Berg, G., & du Plessis, E. *ChatGPT and generative AI: Possibilities for its contribution to lesson planning, critical thinking and openness in teacher education.* Education Sciences, 13(10), 998, 2023. https://doi.org/10.3390/educsci13100998

Warschauer, M., Tseng, W., Yim, S., Webster, T., Jacob, S., Du, Q., & Tate, T. *The affordances and contradictions of AI-generated text for writers and writing pedagogy.* Journal of Second Language Writing, 62, 101057, 2023.

Yeh, H. C. *The synergy of generative AI and inquiry-based learning: Transforming the landscape of English teaching and learning.* Interactive Learning Environments, 1–15, 2024. https://doi.org/10.1080/10494820.2024.2335491

5 Creating an AI-Ready School

Husni Muhamad Rifqi, Tajudeen Alaburo Abdulsalam, and Yahya Fikri

5.1 INTRODUCTION

The effect of Artificial Intelligence (AI) globally is encouraging educational institutions to rethink teaching and learning together with school processes (Song et al., 2023; Southworth et al., 2023). As AI technologies are increasingly embedded in everyday life, schools are transforming beyond the simple adoption of emerging technologies to preparing learners to contribute and live as responsible citizens in an AI society (Kim et al., 2021; Nair, 2024). In the context of the above, the AI ready school model responds to the demand for an integrated approach to educational technology that is systematic, ethical, and grounded in pedagogy, and weaves forth the above description (Somabut et al., 2025).

Developing an AI ready school is a challenging, complex, and systemic undertaking that extends far beyond the simple installation and integration of technologies (Fikri, 2025a, 2025b). Most recent literature shows that the successful integration of AI is shaped and defined by the interconnectedness of the various elements of curriculum design, professional development for teachers, pedagogical and technological tools, school leadership, and the educational goals (Arunfred et al., 2025; Somabut et al., 2025).

The AI readiness of educators constitutes yet another critical dimension. Besides using AI tools, educators have to analyze their pedagogical uses, ethics, and/threat(s). Thus, teacher AI literacy, as argued by El Karafli (2025) encompasses three areas: technical, instructional design, and ethics. The more sophisticated AI and pedagogical methodologies become, the more continuous professional development for educators is needed, aiming, from a pedagogical standpoint, at progressive assimilation—a use that is far more profound than the surface (Keeley, 2024; Ng et al., 2025). Given the AI duality, professional development activities should cover data privacy, algorithmic bias, the digital divide, and other negative sides of AI to encourage responsible and equitable use of AI in education (Mahala & Chauhan, 2025).

AI-ready schools also need strong technological infrastructure and support from both staff and the institution as a whole. Having strong and reliable access to appropriate hardware, software, and networks, plus technical support, forms the foundation for the large-scale implementation of AI (Somabut et al., 2025). However, infrastructure by itself will not be enough. Supportive administrative leadership, as well as effective policies, are needed for the balance between strategic planning, resource allocation, and governance, combined with a good culture of innovativeness and accountability (Arunfred et al., 2025; Fikri, 2025a, 2025b; Fikri et al., 2025). Effective leadership ensures alignment between AI initiatives and institutional goals and the overall educational policies in order to minimize the potential of fragmented initiatives or inequitable distribution of resources.

Student readiness and engagement are likely to be most influential in defining the success of AI integration in schools. Evidence suggests that most AI technologies can improve learning outcomes by personalizing learning and conducting in-process assessments as well as by providing feedback, thus motivating and improving the overall academic performance of the students (Han, 2024). In addition to this, schools are responsible for preparing students to be AI literate, which includes knowing and assessing the potential ethical and moral issues that may accompany the use of AI (Fikri & Rhalma, 2024; Ng et al., 2025).

DOI: 10.1201/9781003767855-5

5.2 A PHASED FRAMEWORK FOR BUILDING AN AI-READY SCHOOL

As AI becomes more prevalent in the educational field, it is necessary for educational institutions to implement a structured and systematic model to use AI, instead of sticking to more fragmented and experimental models. Although quick and ad hoc strategies can assist institutions in adopting new technology, these strategies tend to lead to ethical concerns, a poor use of institutional resources, unequal opportunities given to students, and a lack of optimally AI-supported learning opportunities. These concerns have led to an increased amount of literature that calls for educational institutions to define more aligned and coherent structures to support systematic AI integration in education.

Aside from instructional effectiveness, the structured implementation of educational technology is crucial in considering the potential impact of the AI technology being implemented in education systems. In a school setting, unmediated, ungoverned, and uncalibrated AI systems can create an environment where users' privacy and data protection are ungoverned. In these systems, unregulated bias and poor data governance may remain unaddressed. These concerns are particularly prominent in school systems because the data used in these systems contains sensitive information about the students.

Teacher readiness and professional support highlight the importance of a more defined framework when adopting AI in the classroom. A growing number of studies indicate that the successful adoption of AI in education is primarily dependent on the AI-related knowledge and teaching confidence of the educators concerned (El Karafli, 2025). The more defined implementation frameworks that Yadav (2024) discusses, suggest that educators are given ongoing support as AI tools become integrated into more and more aspects of teaching, rather than having a one-time professional development workshop. This suggests that the more defined implementation supports educators so they do not see the AI technologies as something added on to teaching, and instead see it as integral to teaching.

In relation to these factors, educational institutions are increasingly recommending a phased approach to AI adoption. Phased implementation enables schools to manage their AI integration progressively, synchronizing professional development with each phase, assessing readiness of their infrastructural frameworks, and chronologically embedding organizational structures to ethical barriers (Singh & Thakur, 2024). Furthermore, the adoption of these phases encourages ongoing responsiveness of constituencies involved and adaptive frameworks to governance to balance the value of AI in education against technological determinism (Temimi et al., 2025; Yadav, 2024).

This chapter proposes a five-phase model that will provide the basis for the building of the first AI-ready school. The model will guide the school through the first stages of the responsible, effective, and sustainable integration of AI. The five phases are illustrated in Figure 5.1. The phases of the model are described in a progressive, evolving manner in which feedback and improvement occur. The model consists of the following five phases: (1) Capacity Building and Educator Empowerment, (2) Development of Infrastructure and Data Readiness, (3) Curriculum Redesign and Pedagogical Integration, (4) Ethical Governance and Policy Alignment, and (5) Evaluation, Scaling, and Continuous Improvement. The completion of each of these phases will provide a clear pathway for the school to move from the initial, uninformed integration of AI to a fully equipped, institution-wide preparedness in the integration of AI. A synthesis of the phases is available in Table 5.1, which illustrates the focus of each of the phases, the goals and objectives to be achieved, the major activities to be completed, and the results to be expected. The synthesis provides a structured basis for the school to understand how the phases will provide the school with a means to fully integrate AI in a responsible, ethical, and sustainable manner. Each of the phases will be examined in detail in the following sections.

5.2.1 Phase 1: Capacity Building and Educator Empowerment

Capacity building and educator empowerment are the initial stages of building an AI-ready school (Fikri et al., 2025). Empirical studies repeatedly show that teachers' AI literacy, ethical

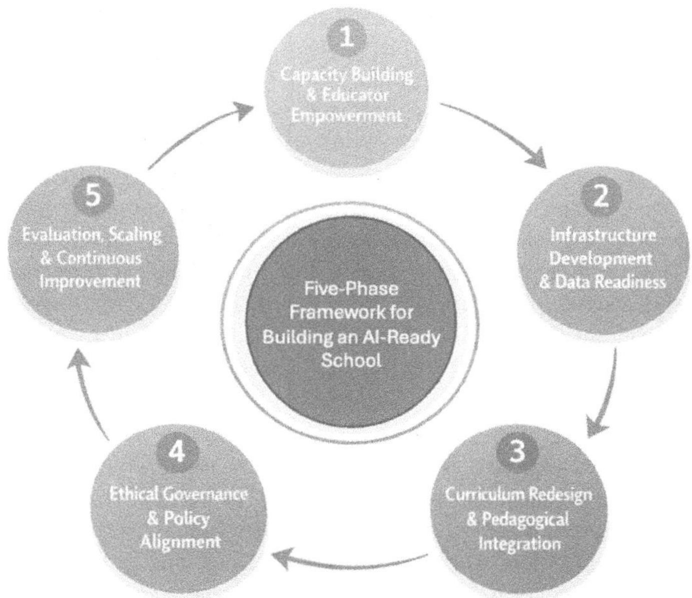

FIGURE 5.1 Five-phase framework for building an AI-ready school.

TABLE 5.1
Phases, Objectives, and Outcomes of the AI-Ready School Framework

Phase	Focus	Key Objectives	Core Actions	Expected Outcomes
Phase 1: Capacity building and educator empowerment	Human readiness	Develop teachers' AI literacy, pedagogical competence, and ethical awareness	• Sustained professional development • Ethical AI literacy training • Support for instructional redesign	Educators empowered as informed, ethical, and confident AI users
Phase 2: Infrastructure development and data readiness	Technological and data foundations	Ensure scalable, equitable, and secure AI implementation	• Investment in hardware, software, and connectivity • Development of high-quality, localized datasets • Data governance and privacy safeguards	Robust, sustainable infrastructure supporting AI-enabled learning
Phase 3: Curriculum redesign and pedagogical integration	Instructional transformation	Embed AI meaningfully into curriculum, pedagogy, and assessment	• Integration of AI across subjects and grade levels • Use of adaptive and data-driven learning tools • Alignment with pedagogical frameworks (e.g., UDL, TPACK)	AI-enhanced, learner-centered, and inclusive teaching and learning
Phase 4: Ethical governance and policy alignment	Institutional governance	Ensure responsible, transparent, and accountable AI use	• Ethical-by-design policies • Data privacy and compliance mechanisms • Multi-stakeholder governance and oversight structures	Trustworthy, equitable, and policy-aligned AI adoption
Phase 5: Evaluation, scaling, and continuous improvement	Sustainability and impact	Monitor effectiveness, scale successful practices, and refine implementation	• Continuous evaluation and feedback • Learning analytics and performance monitoring • Iterative refinement and ethical oversight	Sustainable, scalable, and continuously improving AI-ready school ecosystem

understanding, and teaching confidence are key to determining how integrated or neglected AI tools are in education. Without ongoing and in-depth professional development, AI integration is likely to be surface-level, out of alignment with curricular goals, or ethically problematic.

Educator empowerment in AI-ready schools places significant emphasis on building ethical AI competencies. In a systematic literature review, Tagare et al. (2025) show that effective AI literacy among teachers goes beyond having a grasp of the technology; it also encompasses ethical reasoning, critique of and rationale for the use of AI, and culturally conscious pedagogy. Teachers should be able to critique the workings of an AI system, identify the biases that such systems may contain, and help students to think critically and use AI responsibly. An empirical study in Saudi Arabia also supports this view asserting that teachers' comprehension of AI ethics serves as a strong predictor in gauging and justifying the appropriate selection and usage of AI tools for teaching and learning (Al-Abdullatif, 2025). Overall, this study positions ethical literacy as a significant construct of AI Readiness and defends it as an issue that is far too important to be peripheral.

The evidence suggests that Phase 1 in an AI-ready school environment must include sustained, holistic, and ethically guided professional development. The context must be addressed comprehensively, including in the removal of obstacles and the provision of professional development that is aligned with the triad of technical, pedagogical, and ethics, and in the offering of responsive, sustainable, and evolving professional development to the changing AI systems (Yang et al., 2025). When educators are empowered as active decision-makers, as opposed to mere adopters of technology, schools create the necessary layer of humanity for the later stages of infrastructure development, curriculum incorporation, and governance softening.

5.2.2 Phase 2: Infrastructure Development and Data Readiness

Strategic investment models illustrate how effective the teaching potential of AI really is. Research focused on higher education shows digital literacy, and the level of support and infrastructure of the institution affects the outcomes of the use of AI in education (Ul Haq et al., 2025). Al-Zahrani and Alasmari (2025) also focused on the MENA region, and framed the value of in-country investment in infrastructure and governance, faculty development, and ethics to overcome short-term barriers to implementation and long-term sustainability. While these studies are focused on the higher education sector, they have meaningful implications for the school sector in relation to AI, especially in the areas of institutional capacity and planning.

Data readiness is the second aspect of the infrastructure of AI schools. Educational AI systems are effective only when there are quality and relevant datasets. Jeon et al. (2024) use the principles of sustainability and focus on educational datasets of varying school levels, prioritizing reusability, and developmentally appropriate customization. Such frameworks support long-term AI and data science education goals, decreasing redundancies and inefficiencies in data collection and processing.

Educational datasets that are localized allow for personalized learning that are culturally responsive as well. The use and development of such datasets pose the challenges of privacy issues, algorithmic bias, and the need for large tech investments. Reviews of intelligent tutoring systems that use AI clearly show the data quality, transparency, and ethics as critical components that support sustainable, AI driven learning environments (Lin et al., 2023). Also, empirical studies of the use of generative AI in education indicate that in order to prevent negative outcomes and the misuse of AI, there needs to be clear governance and academic policies around the use of AI (Ramos et al., 2024).

The development of policy and governance frameworks helps support data preparedness and infrastructure development. For AI to be fully integrated into systems, the institutional policies need to address ethics of governance, investment gaps, and collaborative partnerships across sectors (Sharma, 2025). Digital frameworks designed for development in sustainable goals centers on the use of AI for fostering inclusive and equitable education, and warns of exacerbating digital divides when the development of data infrastructure and digital frameworks are unbalanced (Velmurugan et al., 2025). The use of AI technology that is inclusive, and equitable to all learners to bridge the gaps created by poor infrastructure will provide support to those that are most in need (Murari & Parmar, 2025).

The findings suggest that Phase 2 of an AI-ready school must emphasize creating sustainable, scalable, and ethically governed data ecosystems and digital infrastructures. Strategic investment, localized data ecosystem construction, and policy alignment will enable Educational Institutions to build technological infrastructures that will support meaningful curriculum integration and pedagogical innovations in succeeding phases.

5.2.3 Phase 3: Curriculum Redesign and Pedagogical Integration

Integrating redesigning curriculum and pedagogy represents a cornerstone of an AI-ready school, a point that signifies a shift from infrastructural readiness to transformative deep instructional change. At this point, AI is no longer viewed as an exogenous technological tool, but rather as a resource that is integrated into the design of curriculum components, teaching methods, and assessment items. At all levels of education, the research suggests that the use of AI in curriculum design, paired with adaptive learning pathways, fosters a greater degree of learner engagement and inclusivity, provided that the curriculum is integrated into a coherent teaching framework (Çela et al., 2024).

AI integration into the primary education sector has made a marked difference in experimentation with personalized learning and individualized instruction for students with varying needs. Real-time adaptive instruction, continuous feedback, and learning analytics generated by intelligent tutoring systems and large language models (Doğan, 2025) help in tailoring feedback and instruction to individual learners. This technology comes in alignment with the frameworks of Universal Design for Learning by providing learners with varying degrees of ability and disability a range of options with regard to representation, engagement, and expression (Marras et al., 2025). Despite these positive outcomes, studies suggest that the successful integrations of these systems depend on the appropriate use of the constructed AI tools by the educators, often modeled by frameworks such as Technological Pedagogical Subject Matter Knowledge (TPACK) (Marras et al., 2025).

The available literature even though more opportunities exist still identifies significant barriers that are necessary to fix in the process of curriculum innovation. During the various educational contexts, an ethical and privacy issue, including data, algorithm, bias and transparency continues to be problematic (Çela et al., 2024). The literature is more than conclusive that in order to achieve the proposed or anticipated outcomes, the education of the teaching staff has to be prioritized (Doğan, 2025; Marras et al., 2025).

5.2.4 Phase 4: Ethical Governance and Policy Alignment

Integrating AI into teaching, learning, and institutional processes requires ethical governance and policy alignment. These elements form the foundations of an AI-ready school for the future. While previous phases concentrated on human development, infrastructure, and curriculum scope and sequence, Phase 4 focuses on the institutional structures needed to support governance to ensure AI is implemented responsibly, transparently, and with accountability. In the position, without governance structures, the adoption of AI within education can erode the trust of stakeholders, worsen inequities, and create ethical and legal liabilities for educational institutions.

Recent studies highlight the need for multilateral ethics, especially concerning AI educational tools. The AI ethics mountains need to be moved simultaneously, concerning governance, education, privacy, safety, fairness, explainability, and accountability. Ma et al. (2025) highlight the five-dimensional trust model and the necessity of interconnected construct the need for ethical by design through the AI lifecycle. In educational settings, this model mitigates issues of data privacy, algorithmic inequity, and secretive decision-making, which affect the rights of students and educational inequity.

Multi-stakeholder governance models has for such inclusive institutional arrangements the adoption of responsible AI. For example, Cruvinel Júnior et al. (2026) argue that educators, administrators, and policymakers, along with students and developers of the relevant technologies, need to design and work together on the AI policy and practice frameworks for the institution to achieve the desired transparency, accountability, and fairness.

On the institutional level, policies, frameworks, and governance structures act as primary enablers of the articulation of ethical policies. Bhaskar et al. (2025) propose the creation of specialized committees to govern AI ethics, particularly in the fields of education. These committees are expected to focus on, develop, and provide guidance for the creation of policies, as well as frameworks for governance, compliance, and auditing, and respond to the ethical challenges of AI. In educational contexts, these structures could help leadership teams manage AI governance challenges and cultivate an ethical governance culture.

The increasing integration of AI in education raises concerns about data privacy as students' sensitive data is unprotected and data-driven systems are becoming more prevalent. Yan et al. (2025) identify, from an AI ethics framework, the management dimensions of data privacy, data protection, and data sovereignty as foundational. Compliance with the GDPR and FERPA is also a way to advance accountability and provide greater protection for students. However, these regulations mean little without the incorporation of internal mechanisms to practice these principles at the school level.

5.2.5 Phase 5: Evaluation, Scaling, and Continuous Improvement

The final stage of building an AI-ready school involves the assessment of AI activities, scaling up successful strategies, and fostering iterative developments. AI integration can easily splinter if there are no evaluative and feedback structures in place, resulting in an even more misaligned integration than what could have been if there were no initiatives in place in the first place (Arranz-Garcia et al., 2025). Consequently, evaluative frameworks are essential in answering the myriad of questions pertaining to the impact and the levels of AI integration in teaching and learning, and the institution as a whole.

To build up AI initiatives, the institution-wide strategy on the overall goal of the implementation is much more effective than the isolated pilot projects. Well-established AI-assisted learning technologies for K-12 education demonstrate the value of modular and scalable AI-supported learning systems.

The learning analytics, ethical oversight, and continuous improvement through iterative feedback cycles, are positive in and of themselves. For instance, keeping data and processes transparent, as well as often allowing stakeholders to influence the processes iteratively, has shown to improve student engagement and learning outcomes on the AI tools and learning platforms, when regularly evaluated and improved based on outcomes (Kwak, 2025).

5.3 CHALLENGES AND PRACTICAL CONSIDERATIONS

Although AI has the ability to promote innovation in teaching and learning, and the management of schools, in the case of the K-12 sector there are many challenges and constraints to its incorporation in practices. These challenges must be addressed if the inequity problems that the educational systems face are not to be made worse with the incorporation of AI.

A lack of staff professional development and training has often been cited as a lack of the implementation of AI in K-12 education. Educators often state that, in relation to generative AI and data driven systems, they feel unprepared to use AI based tools to assist them in teaching their classes (Arranz-Garcia et al., 2025; Kim, 2025). Additionally, many of the studies report that a lack of professional development of this type often leads educators to underutilize AI and the use of this technology in teaching is often driven by a desire to use the technology instead of focusing on the pedagogy (Singhal et al., 2024).

The identification and analysis of data privacy risks in algorithmic systems and AI reinforce social inequalities and bias in these technologies, and are usually present in most of the research in this area (Crompton et al., 2024). Ethical and privacy issues in AI technologies are even more pertinent in the K-12 setting due to the heightened vulnerability of the learners and the inadequate data governance that most schools are equipped with (Kim, 2025). This results in the slow or, in some cases, complete halt of the implementation of AI systems in education due to the ethical uncertainties that accompany them.

The absence of adequate digital tools, infrastructure, and internet access can impede the inclusive implementation of AI technologies in education (Arranz-Garcia et al., 2025; Pan & Lan, 2023). This is particularly true for schools situated in low resource environments where there is already an absence of funding and technical support to develop the AI-enhanced instrumental learning environments and the additional resources and support that are needed to facilitate the adoption of these systems.

Apart from the structural barriers, negative attitudes and the lack of willingness to change from both teachers and students form practical challenges. This type of resistance is often associated with a lack of AI understanding, insufficient institution frameworks, fear of the problems that may arise, and therefore the lack of clear and open communication and the need for affirmative collaboration and responsive planning change strategies.

The lack of uniform curriculum guidelines in teaching AI in K-12 also poses practical challenges. This is simply because the absence of clearly defined and standardized curricular frameworks leads to inefficiencies in teaching and implementation in a number of schools. This makes it almost impossible to rectify any deficiencies regarding the coherence, continuity, and equity in students' AI learning opportunities (Pan & Lan, 2023). This lack of uniformity also makes it difficult to prepare teachers and evaluate learning outcomes.

5.4 CONCLUSION

Recent studies show that AI integration in K-12 education will be shaped by personalized and adaptive learning systems. Real-time feedback, differentiated learning pathways, and tailored instruction will be helped by intelligent tutoring systems, generative AI and adaptive platforms (Burke & Crompton, 2024; Kim, 2025; Li et al., 2025). This shift focuses on data, learner-centered models that stresses engagement, inclusion, and learning outcomes. These also increase the need for closer pedagogical alignment and supervision to teachers (Suhag & Singh, 2025).

AI-ready schools need to employ more ethical, inclusive, and human-centered strategies when adopting AI. Transparency, fairness, and learner agency in the use of Human-Centered AI and pedagogical intelligence frameworks is advocated by scholars. Governance, clear policies, and a structured approach to the continuing professional development of educators will be required to address the data, algorithm, and tech bias issues. Sustainable AI integration will be directly influenced by the institutions preparedness, leadership and systemic support (Arunfred et al., 2025).

To end, we state that the potential of AI in K-12 education will not be achieved through simplistic technological integration, rather, through thoughtful, multi-level, strategic, and ethical frameworks. Schools will be able to build sustainable ecosystems to be fully prepared for AI by borrowing the applicable steady state adaptive systems from higher education, such as, teaching and learning, institutional effectiveness, and ecosystem administrative systems. Coupled with professional capacity building focusing on inclusive frameworks and iterative design, AI will positively change the face of schooling in a way that will be pedagogically meaningful and socially responsible, and will also change the interaction of the systems (the learning processes, the students, and the AI) for a greater complexity.

REFERENCES

Al-Abdullatif, A. M. (2025). Auditing AI Literacy Competency in K–12 Education: The Role of Awareness, Ethics, Evaluation, and Use in Human–Machine Cooperation. *Systems*, *13*(6), 490. https://doi.org/10.3390/systems13060490

Al-Zahrani, A. M., & Alasmari, T. M. (2025). A comprehensive analysis of AI adoption, implementation strategies, and challenges in higher education across the Middle East and North Africa (MENA) region. *Education and Information Technologies*, *30*(8), 11339–11389. https://doi.org/10.1007/s10639-024-13300-y

Arranz-Garcia, O., del Carmen Romero García, M., & Alonso-Secades, V. (2025). Perceptions, Strategies, and Challenges of Teachers in the Integration of Artificial Intelligence in Primary Education: A Systematic Review. *Journal of Information Technology Education: Research*, *24*, 006. https://doi.org/10.28945/5458

Arunfred, N., Marin, B., Shenbagaraman, V., & Saravanan, P. (2025). *Teacher Readiness and Perceptions Towards AI-Based Smart Teaching Tools.* https://doi.org/10.3233/FAIA251504

Bhaskar, P., Khan, M., & Bhaskar, P. (2025). Establishing an AI Ethics Governance Committee in Higher Education. In *Navigating Modern Digital Communication Ethics and Law* (pp. 105–124). IGI Global Scientific Publishing. https://doi.org/10.4018/979-8-3373-1702-1.ch004

Burke, D., & Crompton, H. (2024). Navigating the Future. In *Artificial Intelligence Applications in K-12* (pp. 215–225). Routledge. https://doi.org/10.4324/9781003440192-13

Çela, E., Fonkam, M. M., Vajjhala, N. R., & Vedishchev, A. (2024). *Artificial Intelligence in Education* (pp. 1–20). https://doi.org/10.4018/979-8-3693-7220-3.ch001

Crompton, H., Jones, M. V., & Burke, D. (2024). Affordances and challenges of artificial intelligence in K-12 education: A systematic review. *Journal of Research on Technology in Education*, *56*(3), 248–268. https://doi.org/10.1080/15391523.2022.2121344

Doğan, A. (2025). *Using Artificial Intelligence in Primary School Mathematics Teaching* (pp. 319–348). https://doi.org/10.4018/979-8-3373-1998-8.ch011

El Karafli, K. (2025). *Empowering Educators With AI* (pp. 369–400). https://doi.org/10.4018/979-8-3373-1827-1.ch014

Fikri, Y. (2025a). The Moroccan evidence through an ARDL approach to the education, renewable energy consumption, and economic growth nexus. *International Journal of Energy Economics and Policy*, *15*(6), 312.

Fikri, Y. (2025b). What about poverty in early childhood education? *Sciences (UKRJAHSS)*, *1*(6), 238–244, https://doi.org/10.5281/zenodo.16788010

Fikri, Y., & Rhalma, M. (2024). Artificial Intelligence (AI) in early childhood education (ECE): Do effects and interactions matter. *International Journal of Religion: Politics, Sociology, Culture*, *5*(11), 7536–7545.

Fikri, Y., Talaat, R., Shaheen, A., Hassan, A., & Khataan, A. (2025). Through the ARDL approach: Is there a nexus between renewable energy consumption, economic growth, and foreign direct investment in the Moroccan context? *Sustainability*, *17*(21), 9762.

Han, Y. (2024). *Research on the Influence of AI Technology on University Learning Effect and Optimization Strategy* (pp. 208–218). https://doi.org/10.1007/978-3-031-69457-8_20

Jeon, I.-S., Kim, S.-Y., & Kang, S.-J. (2024). Developing standards for educational datasets by school level: A framework for sustainable K-12 education. *Sustainability*, *16*(12), 4954. https://doi.org/10.3390/su16124954

Júnior, C., Héctor Ascama, L., & Marques da Silva, O. M. (2026). *AI Ethics in Higher Education: A Review of Ethical Challenges* (pp. 192–203). https://doi.org/10.1007/978-3-032-01429-0_17

Keeley, K. L. (2024). *AI Applications and Strategies in Teacher Education* (K. L. Keeley, Ed.). IGI Global. https://doi.org/10.4018/979-8-3693-5443-8

Kim, J. (2025). Perceptions and preparedness of K-12 educators in adopting generative AI. *Research in Learning Technology*, *33*. https://doi.org/10.25304/rlt.v33.3448

Kim, S., Jang, Y., Kim, W., Choi, S., Jung, H., Kim, S., & Kim, H. (2021). Why and what to teach: AI curriculum for elementary school. *Proceedings of the AAAI Conference on Artificial Intelligence*, *35*(17), 15569–15576. https://doi.org/10.1609/aaai.v35i17.17833

Kwak, M. (2025). The effectiveness of AI-driven tools in improving student learning outcomes compared to traditional methods. *Issues In Information Systems*. https://doi.org/10.48009/4_iis_2025_120

Li, D., Osman, S., Alhassora, N., Amantha Kumar, A., & Kartini Said Husain, J., S. (2025). Empowering K-12 mathematics teachers with artificial intelligence: A systematic review of insights, applications, and challenges. *IEEE Access*, *13*, 117198–117209. https://doi.org/10.1109/ACCESS.2025.3585784

Lin, C.-C., Huang, A. Y. Q., & Lu, O. H. T. (2023). Artificial intelligence in intelligent tutoring systems toward sustainable education: A systematic review. *Smart Learning Environments*, *10*(1), 41. https://doi.org/10.1186/s40561-023-00260-y

Ma, Y., Li, X., Hu, S., Liu, S., & Cheong, K. H. (2025). Trustworthy AI in education: Framework, cases, and governance strategies. *Innovation and Emerging Technologies*, *12*. https://doi.org/10.1142/S2737599425500264

Mahala, A., & Chauhan, B. (2025). How can the AI-generated curriculum gap in teacher education be addressed? *TechTrends*, *69*(6), 1158–1167. https://doi.org/10.1007/s11528-025-01102-z

Marras, A., Fornara, F., & Pasqualotto, A. (2025). Harnessing AI for Inclusive Teaching: Professional Development Strategies for Early Educators. In *AI in Early Education* (pp. 267–290). Wiley. https://doi.org/10.1002/9781394352821.ch16

Murari, U. K., & Parmar, H. (2025). *Ethical Integration of Artificial Intelligence in Inclusive Education* (pp. 439–470). https://doi.org/10.4018/979-8-3373-2262-9.ch015

Nair, H. B. (2024). AI Preparedness for Staff and Educators: Equipping the Next Beneration: A Pre-Service Teacher Education Programme in Integrating AI Literacy Across the Curriculum. In *Effective Practices in AI Literacy Education: Case Studies and Reflections* (pp. 30–40). Emerald Group Publishing Ltd.

Ng, D. T. K., Chan, E. K. C., & Lo, C. K. (2025). Opportunities, challenges and school strategies for integrating generative AI in education. *Computers and Education: Artificial Intelligence*, *8*, 100373. https://doi.org/10.1016/j.caeai.2025.100373

Pan, J., & Lan, X. (2023). *Research on the Current Situation and Implementation Strategies of Artificial Intelligence (AI) Education in K-12 Schools* (pp. 519–528). https://doi.org/10.1007/978-3-031-29016-9_44

Ramos, R. F., De Angel, R. M., Ruetas, A. P., Enrile, J. C., Calimbo, A. L., & Vargas, P. C. (2024). Impact Assessment of ChatGPT and AI Technologies Integration in Student Learning: An Analysis for Academic Policy Formulation. *2024 6th International Workshop on Artificial Intelligence and Education (WAIE)*, 87–92. https://doi.org/10.1109/WAIE63876.2024.00023

Sharma, A. (2025). Revolutionizing Education: Artificial Intelligence-Powered Learning in Business Education. In *Transforming Business Education Through Artificial Intelligence* (pp. 423–442). Apple Academic Press. https://doi.org/10.1201/9781998511471-21

Singh, G., & Thakur, A. (2024). *AI in Education: Ethical Challenges and Opportunities* (pp. 18–38). https://doi.org/10.4018/979-8-3693-2964-1.ch002

Singhal, A., Singh, R., & Jansi, K. R. (2024). AI-K12 Connect: Shaping Tomorrow's Scholars. *2024 2nd International Conference on Networking and Communications (ICNWC)*, 1–4. https://doi.org/10.1109/ICNWC60771.2024.10537401

Somabut, A., Tuamsuk, K., Lowatcharin, G., Traiyarach, S., & Kwangmuang, P. (2025). Preparing for the AI era: Science teachers' readiness and professional development needs for generative AI integration in secondary education. *Social Sciences & Humanities Open*, *12*, 102259. https://doi.org/10.1016/j.ssaho.2025.102259

Song, J., Yu, J., Yan, L., Zhang, L., Liu, B., Zhang, Y., & Lu, Y. (2023). Develop AI teaching and learning resources for compulsory education in China. *Proceedings of the AAAI Conference on Artificial Intelligence*, *37*(13), 16033–16039. https://doi.org/10.1609/aaai.v37i13.26904

Southworth, J., Migliaccio, K., Glover, J., Glover, J., Reed, D., McCarty, C., Brendemuhl, J., & Thomas, A. (2023). Developing a model for AI across the curriculum: Transforming the higher education landscape via innovation in AI literacy. *Computers and Education: Artificial Intelligence*, *4*, 100127. https://doi.org/10.1016/j.caeai.2023.100127

Suhag, N., & Singh, A. (2025). *Blended Learning in the Age of Artificial Intelligence* (pp. 29–52). https://doi.org/10.4018/979-8-3373-3815-6.ch002

Tagare, D., Karki, T., & Yu, W. (2025). K-12 teachers' ethical competencies for AI literacy: Insights from a systematic literature review. *Computers & Education*, *239*, 105435. https://doi.org/10.1016/j.compedu.2025.105435

Temimi, A., Rebai, R., Aldamen, H., Souki, K., & Amiri, K. (2025). Education at a crossroads: Will AI fix learning or break it? *Management & Sustainability: An Arab Review*, 1–25. https://doi.org/10.1108/MSAR-04-2025-0118

ul Haq, F., Asim, M., Suki, N. M., Zakaria, N., & Hussain, S. (2025). AI adoption and educational effectiveness in emerging higher education Institutions: The moderating role of digital literacy and institutional support. *Journal of Information & Knowledge Management.* https://doi.org/10.1142/S021964922550090X

Velmurugan, R., Kalimuthu, M., Bhuvaneswari, R., & Abey, J. (2025). Digital Strategies for Supporting Education Aligned With the Sustainable Development Goals (SDGs). In *Digital Transformation and Sustainability in Higher Education* (pp. 303–332). IGI Global Scientific Publishing. https://doi.org/10.4018/979-8-3373-5077-6.ch012

Yadav, S. (2024). *Empowering Educators for the Artificial Intelligence (AI) Era* (pp. 1–28). https://doi.org/10.4018/979-8-3693-9806-7.ch001

Yan, Y., Liu, H., Zhang, H., Chau, T., & Li, J. (2025). Designing a generalist education AI framework for multimodal learning and ethical data governance. *Applied Sciences, 15*(14), 7758. https://doi.org/10.3390/app15147758

Yang, H., Lee, W., & Kim, J. (2025). Identification of key factors influencing Teachers' self-perceived AI literacy: An XGBoost and SHAP-based approach. *Applied Sciences, 15*(8), 4433. https://doi.org/10.3390/app15084433

6 From Policy to Practice

Exploring AI Integration in Higher Education

Meet Purohit and Bhavna Sharma

6.1 INTRODUCTION

Artificial intelligence (AI) is the term for machines that carry out tasks requiring human intelligence, including learning, reasoning and adapting. It is now widely used in higher education around the globe; in fact, one study finds that AI tools are being used more and can improve student learning, often thanks to adaptive platforms and robots. In India, the National Education Policy of 2020 encourages a significant role for technology in improving teaching and learning (Xie et al., 2021). In the Higher Education Institutions, the role of AI is now being studied in teaching, administrative work, and student support. It is indicated in recent studies that universities using AI tools, such as hybrid course recommendation systems, noticeably increase student interest and let students make their own decisions about their courses', meaning AI has practical use in support of students. Still, experts explain that despite AI spreading throughout education, educators have difficulty adapting due to rapid changes. AI impacts every part of teaching and studying, institutions should consider it from every angle. One model demonstrates that AI can support learners alone, facilitate interactions among team members and help build knowledge at the institution. Practically speaking, AI affects teaching approaches, what is taught, how students learn and how instruction is evaluated. In the case of individual education, intelligent tutors and customised suggestion algorithms show examples of AI, with AI analyses and teamwork programmes focusing on collaborative accomplishment for a group and administrative management (Gibson et al., 2023). Seeing these parts of an AI system reminds us to study it independently and as part of the curriculum, teaching, student support and assessment. As a result, this study looks at how AI is helpful and problematic and has possible policy impacts within different parts of educational institutions. We discuss how AI is used for innovative teaching methods, interactive learning materials, student focus, engagement, assessment, and institution running. We examine Indian and international experiences following the NEP 2020 focus on technology and review research evidence from experts. The study covers the main opportunities and difficulties higher education encounters by relating AI to the overall set of institutional elements.

6.2 LITERATURE REVIEW

Crompton and Burke (2023) reported that publications on AI in higher education roughly doubled in 2021–2022 compared to prior years. Their study from 2016 to 2022 identified principal "usage codes" such as automated assessment, predictive analytics, AI assistants, intelligent tutoring systems and learner management, with a notable shift from U.S.-centric research to leading contributions from China and other regions. By 2022, initial work noted AI's promise in personalising instruction and administrative efficiency, though formal evaluation frameworks were still nascent. In 2023, empirical and review studies emerged rapidly. Crompton and Burke (2023) Furthermore, most AIED research focused on students (72%), especially in language learning, underscoring adaptive and personalised learning as dominant themes. Chan and Hu (2023) documented generally

DOI: 10.1201/9781003767855-6

positive attitudes among Hong Kong university students towards generative AI tools (e.g. chatbots for writing support), noting enhanced personalisation and writing assistance but concerns about accuracy, privacy, and academic integrity. Gligorea et al. (2023) concluded that AI-driven adaptive learning systems "significantly" improved personalisation, engagement, and academic performance in e-learning, while flagging challenges like data privacy, bias and system complexity. Bond et al. (2024) confirmed that adaptive systems and personalisation dominated AI in higher ed research, but called for stronger ethical and contextual rigour in future work. Kalniņa et al. (2024) reported that pre-service teachers held mixed views of AI as a pedagogical tool; fewer than half used AI in practice, expressing concerns about reduced critical thinking and student agency. They stressed the need for AI literacy, formal policies and assessment redesign to realise AI's benefits without undermining learning.

In 2024, the focus shifted to generative AI and its effects. Yusuf et al. (2024) conducted a worldwide survey of over 1,200 higher education stakeholders across 76 countries and found widespread awareness and intended use of GenAI (primarily for information retrieval and text paraphrasing) alongside significant worries about academic dishonesty and bias. Their multicultural analysis revealed strong correlations between cultural context and attitudes towards AI ethics. Xia et al. (2024) found that GenAI's penetration in assessments requires fundamental changes: they conclude that assessment should be re-designed to cultivate self-regulated learning and integrity, recommending extensive teacher training in AI and digital literacy, strengthened trust in human versus AI evaluation, and holistic, innovative teaching aligned with these goals. A study surveyed Saudi higher education faculty and administrators and similarly confirmed optimism about AI's ability to enhance teaching, learning and administrative work, but emphasised the urgent need for ethics guidelines, data privacy protections and fairness safeguards as AI tools are adopted. Merino-Campos (2025) systematically reviewed AI in personalised learning and concluded that AI-tailored content "significantly optimises educational outcomes" by customising feedback and pathways, but also consistently raised ethical and privacy concerns along with the necessity for instructor training. The studies reinforced global findings: AI can boost engagement and efficiency, but persistent challenges from academic integrity to equity must be managed with policy and pedagogy changes.

Mulaudzi and Hamilton (2025) found that instructors responding to AI use in teaching emphasise integrity measures like AI detection tools, transparent AI-use policies, redesigned assessments and fostering students' critical thinking; they advocate comprehensive training for faculty and students on ethical AI use. Ren and Wu (2025) identified core teaching competencies for AI integration – including AI literacy, AI-enhanced pedagogy design, and experience-centric instruction – while noting integration hurdles such as novelty effects, the digital divide and technical issues, all echoing earlier ethical worries. Ocen et al. (2025) discussed that AI is "revolutionising" higher education by automating tasks from admissions to academic records and financial aid, freeing staff for strategic work, and enabling personalised learning pathways for students. Their review underscores that strategic leadership and governance need to align AI deployment with institutional mission. Liang et al. (2025) found that AI's promise includes content generation, instructor workload reduction, and automated evaluation. However, meaningful gains will require addressing staff training and systemic challenges. Throughout this period, ethical issues remain prominent. Bittle and El-Gayar (2025) reported that while GenAI can increase learner engagement and efficiency, it dramatically heightens opportunities for plagiarism and misconduct, thus demanding improved digital literacy and AI-detection strategies.

In many settings – North America, Europe, Asia, the Middle East (Yusuf et al., 2024; Al-Zahrani & Alasmari, 2024; Reina Marín et al., 2025) AI is seen as a tool that can support both personalised learning and improved administration. However, the paper draws attention to ongoing issues such as data privacy, how AI affects algorithms, changes needed in assessment practices, and the development of faculty skills in AI. As research continued, scholars began by setting goals to integrate AI, shifted to studying stakeholder reactions, now focused on signing policies, and recently examined how AI can be implemented on a large scale. These studies picture a worldwide situation in which universities see AI as an important force that can change how they operate, even as those in charge review its ethical and practical side effects.

6.3 OPPORTUNITIES AND CHALLENGES

Since AI powers adaptive learning systems, they can adapt lessons, the speed and feedback to each person, helping them to improve engagement, stay motivated and perform well. Thanks to these tools, students feel more certain about their understanding and find it easier to remember what they have learned in online and large-enrolment classes (Yaseen et al., 2025). AI can also help institutions work more efficiently: for instance, intelligent data systems can improve how courses are scheduled, how students enrol and how they receive student support, while reducing expenses and faculty tasks, freeing up time for staff (Sposato, 2025). AI in teaching helps target a variety of learning needs. It provides students with helpful resources and guidance to match what they learn in class and to help them progress at their own pace, especially for those in remote or less well-equipped areas and provides content in text, video and interactive formats to match learners' preferences (Bond et al., 2024). Digital grading tools for essays and assignments, along with fast feedback, make it possible to identify challenges for many students early and release teachers from routine grading so they can help individual learners (Crompton & Burke, 2023). Such improvements indicate that AI could democratise learning worldwide, thanks to Indian institutions piloting new adaptive systems to help students and instructors (Yaseen et al., 2025). However, fast adoption of AI creates big worries about academic honesty: many teachers and students in various countries are concerned that generative AI encourages plagiarism and cheating in schoolwork, showing up in survey responses from 76 countries (Yusuf et al., 2024). There are still important issues regarding ethics and operation: having trained AI with biased data can increase bias in treatment, and collecting a massive amount of student data may harm their privacy. Therefore, without proper policies and faculty education in AI, these problems could widen the gap in educational opportunities offered. Moreover, because many developing countries (including India) cannot yet manage these risks, the need for worldwide standards and capacity-building becomes clear (Merino-Campos, 2025).

6.4 VIEWPOINTS OF FACULTY

Do not be surprised if faculty express some excitement and worry about AI. As an example, researchers from the U.S. discovered that how helpful people believe generative AI to be is a much more decisive factor influencing their desire to adopt it than how easy it seems to them (Shata & Hartley, 2025). According to Spanish studies, although teachers and administrators were interested in GenAI, they were usually concerned about students' learning and ChatGPT misuse, so further technical training and support are needed to solve these concerns (Verano-Tacoronte et al., 2025). When you look at educators' social media posts, about 40% are positive (focused on trust or joy), 51% are neutral, and only 9% are critical of ChatGPT, suggesting neutral or mostly positive views towards AI within the faculty body (Mamo et al., 2024). Similarly, an Indian research study finds that teachers' AI self-esteem, how effective they think AI is, their institution's encouragement and planning to use AI are all linked, stressing the need for professional help and support (Svihus, 2024).

6.5 INSTITUTIONAL INITIATIVES

Communities in universities are organising formal systems for learning about and managing AI. The University of Florida runs an "AI Across the Curriculum" programme to ensure faculty in multiple disciplines feel comfortable using AI (Ayyoub et al., 2025). Most 14 U.S. universities have issued GenAI suggestions and set up committees to handle responsible use (Wu et al., 2024). They are similar to experts' findings who say AI is a key method to improve how universities organise learning and their administration (Sahar & Munawaroh, 2025). According to these views, international organisations and governments, including India, are taking purposeful measures to adopt AI.

6.6 AI's USAGE IN THE INSTITUTIONS

Personal learning through adaptive systems has become popular in higher education, largely thanks to AI. When Arizona State used Knewton's Alta platform, which analysed quiz and homework results, introducing personalised practice problems for each student, the class pass rates spiked above those in typical classes (Strielkowski et al., 2024). Generally, AI systems for education and course advisors can personally guide students; students who used AI dashboards and tutoring systems said they were more engaged and motivated because they immediately saw how they were progressing and were given advice on what to do next (Ramaswami et al., 2023). Plus, AI analytics allow teachers to monitor students' learning: a study has revealed that students increased their study efforts and got more involved in class after getting visual insights into their performance using an AI dashboard (Ramaswami et al., 2023). Machine learning models in classrooms help teachers detect students who are in danger of not keeping up. Because AI analysed data from student participation and limited site use, the distance learning programme arranged Indian students into groups, seeing them as [not] likely to succeed, and the groups were evaluated correctly in 85% of cases to provide timely assistance for those in need (Liang et al., 2025). These tools guide teachers to focus on struggling students and, if necessary, use additional resources to make learning available to a broader range of students. AI is drastically changing how we conduct assessments. It is possible to use advanced automatic models to assess student essays and question answers. Real university paper responses were evaluated by a large language model (ChatGPT-3.5) without being trained, and its scores were as close to human graders' marks in 70% of cases (Flodén, 2024). So it looks like AI can tackle open-ended questions that teachers have often worried were too challenging to measure accurately. In addition, researchers suggest that AI graders tend not to give very high or low scores, preferring to group grades close to the middle. Research also revealed that AI gave fewer teachers A+ or F scores than human instructors (Flodén, 2024). Still, this compression can be improved by adding more detailed guidance and examples. Besides major tests, AI offers different support: students can try quizzes, receive feedback on their writing and work with virtual teaching tools. These systems take over the basic grading work and encourage students to correct their mistakes quickly. All this puts the assessment process squarely into the regular classroom, not just at testing and exam periods. AI must be integrated throughout school training and teaching, which is still tricky. Discussions are happening now on teaching AI – some favour standalone AI lessons, but others want AI elements added to current courses. When we surveyed journalism teachers, the choices included providing AI as a standalone course or gradually adding examples about AI in media classes (Liang et al., 2025). Whatever the teaching method, most institutions believe it is important to incorporate algorithmic literacy, data privacy and responsible AI into their curriculum. Educators often use AI, such as chatbot tutors or automated content, to improve their courses and lesson plans. As a result, AI leads to updated lessons and new methods involving people and AI.

6.7 AI-DRIVEN CONTENT

AI makes it possible for students to get learning materials that match their needs, whether they are text, multimedia or simulations and automatically (Rismanchian & Doroudi, 2025). Large language models and image generators can, on their own, build quizzes, give examples and produce visual material that follows the goals of the course (Tang et al., 2025). Using student data, AI-powered tools change the learning design and content to keep challenges effective and fit the students' growth (Rismanchian & Doroudi, 2025). Some teachers use digital tutors called AI "teachable agents" or chatbots to supply personalised hints and illustrations at any time (Tang et al., 2025). Generative-AI-aided learning helped students show greater satisfaction and engagement with their studies, but they often needed more support from their instructors before reaching higher learning outcomes (Tang et al., 2025). AI opens new ways to teach and learn, and science agrees that good education needs instructors to oversee and set pedagogical strategies (Tang et al., 2025).

6.8 CONCLUSION

AI in higher education could allow everyone to receive personalised, scaled learning, but there is a concern that over automating could make things impersonal in classrooms. Internationally, reviews show that AI helps by customising lessons and guiding learning, which improves academic performance for students from different places. Nevertheless, their readiness primarily determines how much faculty use AI-driven technology. Findings from research stress the need for programmes that help professors become AI literate and develop trust in AI. Robust research has found AI is being put to use in grading, identifying why students drop out, tutoring support and providing personalised content, indicating the technology influences various fields of education and administration. These results prove that AI is transforming how teaching, tests and curriculum are built at the same time. As an example, there is global experimentation with AI in the classroom using generative systems and intelligent tutors, which both change delivery methods and support for students and lead to important questions about instructional design and evaluation. Such changes in universities are important, as current educational reform agendas in many countries (like India's NEP 2020) chiefly stress the use of digital and AI tools in teaching, urging universities to make sure their systems and support mechanisms are up to date. Even so, many difficulties persist: organisations may not have the facilities and specialists that could hinder their ability to use AI fairly and without proper guidelines, and good AI intentions might backfire by increasing bias or leaving privacy unprotected. Thus, the paper concludes that leaders in higher education institutions should manage the use of AI sensibly, encourage various fields to work together and base their approaches on research findings, so that any improvement from technology leads to suitable and honest learning outcomes.

6.9 IMPLICATIONS AND FUTURE SCOPE

As things stand, higher education institutions should ensure educators know how to handle, use and assess AI in education at its most fundamental level. As well as updating resources, this means adding learning support: research in classrooms reveals that many students may not grasp the needed knowledge to critique AI and therefore should be supported by learning about AI along with their primary discipline. Fortunately, trials have demonstrated that presenting material through AI can boost college students' grades: it was discovered in a meta-analysis that well-designed AI tools lead to medium-sized improvements. Research has shown that using AI to support learning has obvious benefits – a chatbot using ChatGPT technology helped secondary students learn more science, improved their attention and motivation, and reduced their anxiety. This builds on students' responses: many undergraduates rapidly accept AI help and notice improvements in critical thinking, showing that strictly banning AI is ineffective and that education should urge students to be responsible with AI. Even so, using AI everywhere creates several clear issues. In South Asian universities, researchers explain, many challenges can be seen: there is no clear policy, people are not sure about AI's reliability, faculty are not well-matched for the task and cultural issues get in the way. To make the most of AI's potential and reach the goals stated in India's NEP 2020 for inclusive quality education, scholars suggest teamwork by many stakeholders. Recent research suggests that people from different sectors – education, technology and policymaking – should work as teams, create strong AI practices and check AI's impact regularly. To conclude, the new literature supports a plan for universities to develop good governance, prepare for growth, enhance teaching skills, collaborate with faculty and build AI-based courses centred on fairness and respect. Higher education can promote better learning outcomes everywhere in India by combining innovation and proven risk management strategies.

REFERENCES

Ayyoub, A. M., Khlaif, Z. N., Shamali, M., Eideh, B. A., Assali, A., Hattab, M. K., Barham, K. A., & Bsharat, T. R. K. (2025). Advancing higher education with GenAI: Factors influencing educator AI literacy. Frontiers in Education, 10. https://doi.org/10.3389/feduc.2025.1530721

Bittle, K., & El-Gayar, O. (2025). Generative AI and academic integrity in higher education: A systematic review and research agenda. Information, 16(4), 296. https://doi.org/10.3390/info16040296

Bond, M., Khosravi, H., de Laat, M., Bergdahl, N., Negrea, V., Oxley, E., Pham, P., Chong, S. W., & Siemens, G. (2024). A meta-systematic review of artificial intelligence in higher education: A call for increased ethics, collaboration, and rigour. International Journal of Educational Technology in Higher Education, 21, Article 4. https://doi.org/10.1186/s41239-023-00436-z

Chan, C. K. Y., & Hu, W. (2023). Students' voices on generative AI: Perceptions, benefits, and challenges in higher education. International Journal of Educational Technology in Higher Education, 20, Article 43. https://doi.org/10.1186/s41239-023-00411-8

Crompton, H., & Burke, D. (2023). Artificial intelligence in higher education: The state of the field. International Journal of Educational Technology in Higher Education, 20, Article 22. https://doi.org/10.1186/s41239-023-00392-8

Flodén, J. (2024). Grading exams using large language models: A comparison between human and AI grading of exams in higher education using ChatGPT. British Educational Research Journal. https://doi.org/10.1002/berj.4069

Gibson, D., Kovanovic, V., Ifenthaler, D., Dexter, S., & Feng, S. (2023). Learning theories for artificial intelligence promoting learning processes. British Journal of Educational Technology, 54(5), 1125–1146. https://doi.org/10.1111/bjet.13341

Gligorea, I., Cioca, M., Oancea, R., Gorski, A.-T., Gorski, H., & Tudorache, P. (2023). Adaptive learning using artificial intelligence in e-learning: A literature review. Education Sciences, 13(12), 1216. https://doi.org/10.3390/educsci13121216

Kalniņa, D., Nīmante, D., & Baranova, S. (2024). Artificial intelligence for higher education: Benefits and challenges for pre-service teachers. Frontiers in Education, 9, 1501819. https://doi.org/10.3389/feduc.2024.1501819

Liang, J., Stephens, J. M., & Brown, G. T. L. (2025). A systematic review of the early impact of artificial intelligence on higher education curriculum, instruction, and assessment. Frontiers in Education, 10, Article 1522841. https://doi.org/10.3389/feduc.2025.1522841

Mamo, Y., Crompton, H., Burke, D., & Nickel, C. (2024). Higher Education Faculty Perceptions of CHATGPT and the Influencing Factors: A Sentiment Analysis of X. TechTrends, 68(3), 520–534. https://doi.org/10.1007/s11528-024-00954-1

Merino-Campos, C. (2025). The impact of artificial intelligence on personalized learning in higher education: A systematic review. Trends in Higher Education, 4(2), 17. https://doi.org/10.3390/higheredu4020017

Mulaudzi, L. V., & Hamilton, J. (2025). Lecturers' perspective on the role of AI in personalized learning: Benefits, challenges, and ethical considerations in higher education. Journal of Academic Ethics, 2025, Article 25. https://doi.org/10.1007/s10805-025-09615-1

Ocen, S., Elasu, J., Manjeri Aarakit, S., Olupot, C., Haruna, A., & Nyambura, H. (2025). Artificial intelligence in higher education institutions: Review of innovations, opportunities and challenges. Frontiers in Education, 10, 1530247. https://doi.org/10.3389/feduc.2025.1530247

Ramaswami, G., Susnjak, T., & Mathrani, A. (2023). Effectiveness of a learning analytics dashboard for increasing student engagement levels. Journal of Learning Analytics, 10(3), 115–134. https://doi.org/10.18608/jla.2023.7935

Reina Marín, Y. R., Cruz Caro, O., Maicelo Rubio, Y. C., Alva Tuesta, J. N., Sánchez Bardales, E., Carrasco Rituay, A. M., & Chávez Santos, R. (2025). Artificial intelligence as a teaching tool in university education. Frontiers in Education, 10, 1578451. https://doi.org/10.3389/feduc.2025.1578451

Ren, X., & Wu, M.-L. (2025). Examining teaching competencies and challenges while integrating artificial intelligence in higher education. TechTrends. https://doi.org/10.1007/s11528-025-01055-3

Rismanchian, S., & Doroudi, S. (2025). The evolution of research on AI and education across four decades: Insights from the AIxEd framework. International Journal of Artificial Intelligence in Education. https://doi.org/10.1007/s40593-025-00483-9

Sahar, R., & Munawaroh, M. (2025). Artificial intelligence in higher education with bibliometric and content analysis for future research agenda. Discover Sustainability, 6(1). https://doi.org/10.1007/s43621-025-01086-z

Shata, A., & Hartley, K. (2025). Artificial intelligence and communication technologies in academia: Faculty perceptions and the adoption of generative AI. International Journal of Educational Technology in Higher Education, 22(1). https://doi.org/10.1186/s41239-025-00511-7

Sposato, M. (2025). Artificial intelligence in educational leadership: A comprehensive taxonomy and future directions. International Journal of Educational Technology in Higher Education, 22, Article 20. https://doi.org/10.1186/s41239-025-00517-1

Strielkowski, W., Grebennikova, V., Lisovskiy, A., Rakhimova, G., & Vasileva, T. (2024). AI-driven adaptive learning for sustainable educational transformation. Sustainable Development. https://doi.org/10.1002/sd.3221

Svihus, C. L. (2024). Online teaching in higher education during the COVID-19 pandemic. Education and Information Technology, 29, 3175–3193. https://doi.org/10.1007/s10639-023-11971-7

Tang, Q., Deng, W., Huang, Y., Wang, S., & Zhang, H. (2025). Can generative artificial intelligence be a good teaching assistant?—An empirical analysis based on generative AI-assisted teaching. Journal of Computer Assisted Learning, 41(3). https://doi.org/10.1111/jcal.70027

Verano-Tacoronte, D., Bolívar-Cruz, A., & Sosa-Cabrera, S. (2025). Are university teachers ready for generative artificial intelligence? Unpacking faculty anxiety in the ChatGPT era. Education and Information Technologies. https://doi.org/10.1007/s10639-025-13585-7

Wu, C., Zhang, H., & Carroll, J. M. (2024). AI governance in higher education: Case studies of guidance at big ten universities. Future Internet, 16(10), 354. https://doi.org/10.3390/fi16100354

Xia, Q., Weng, X., Ouyang, F., Lin, T. J., & Chiu, T. K. F. (2024). A scoping review on how generative artificial intelligence transforms assessment in higher education. International Journal of Educational Technology in Higher Education, 21, Article 40. https://doi.org/10.1186/s41239-024-00468-z

Xie, H., Hwang, G.-J., & Wong, T.-L. (2021). From conventional AI to modern AI in education: Re-examining AI and analytics techniques for teaching and learning. Educational Technology & Society, 24(2021). https://www.j-ets.net/collection/published-issues/24_3

Yaseen, H., Mohammad, A. S., Ashal, N., Abusaimeh, H., Ali, A., & Sharabati, A. A. (2025). The impact of adaptive learning technologies, personalized feedback, and interactive AI tools on student engagement: The moderating role of digital literacy. Sustainability, 17(3), 1133. https://doi.org/10.3390/su17031133

Yusuf, A., Pervin, N., & Román-González, M. (2024). Generative AI and the future of higher education: A threat to academic integrity or reformation? Evidence from multicultural perspectives. International Journal of Educational Technology in Higher Education, 21, Article 21. https://doi.org/10.1186/s41239-024-00453-6

7 AI and IoT-Enabled Smart Classrooms

Redefining Teaching–Learning Processes in Higher Education

Shyamasundar Tripathy, José G. Vargas-Hernández, and Islam Elgammal

7.1 INTRODUCTION

The pace of rapid development in technology is transforming the higher education domain, mainly led by digital, artificial intelligence (AI) and Internet of Things (IoT). Driven by such technology diffusion, smart environments and technologies have promoted the creation of smart classrooms in educational institutions for enhancing teaching efficiency, students' engagement and learning achievement (Hashim et al., 2022).

AI in education also supports personalised learning by means of adaptive content delivery, automated grading, predictive analytics and just-in-time feedback to tailor instruction to students' diverse learning needs (Sundari, Penthala, and Nayyar 2024). IoT acts as an adjunct to AI, as it is useful in ensuring effortless interconnections between physical and digital learning resources which allow the real-time monitoring of classroom activities, student attendance and environmental situations (Kumar et al., 2025; Kurni and Srinivasa, 2025a, 2025b). Composed, AI and IoT form an intelligent learning ecosystem for near real-time assessment, collaborative learning and evidence-informed teaching decisions (Wangoo and Reddy, 2020).

Internationally, smart classroom approaches are being implemented in higher education as a response to challenges including student diversity, the demand for flexible learning and to improve academic performance and institutional effectiveness (Acharya 2019; Müller, Mildenberger, and Steingruber, 2023). But it is not all sunshine and roses when AI gets added to IoT for education (Ahuja and Bala, 2021).

In the context of India, various steps like the launch of Digital India, National Education Policy (NEP) 2020 and scaling up EdTech platforms have hastened the implementation of digital technologies in higher education (Ahuja and Bala, 2021; Nagaraju and Gorabal, 2025). Some states like Andhra Pradesh have heavily invested in digital classrooms, smart boards and online educational platforms throughout the public and private sectors. However, evidence on how AI- and IoT-based smart classrooms are actually affecting the teaching–learning process at the institutional level is scant, especially in terms of primary data (Huang, 2024; Marfo et al., 2025). This lacuna is filled by the present study through an analysis of AI and IoT-facilitated smart classrooms in re-shaping teaching–learning processes in Higher Education Institutions (HEIs) of Andhra Pradesh, India. Through an examination of faculty and student experiences, the research contributes a micro-level perspective on patterns of adoption, perceived benefits and implementation challenges. This kind of evidence is essential to enable the decision-making processes in institutions, design policies and strategies and sustainable implementations of smart-classroom technologies within higher education systems.

 DOI: 10.1201/9781003767855-7

7.2 LITERATURE REVIEW

The fusion of AI and IoT has become one of the characteristics of digital transformation in higher education (Koong et al., 2023). Together, these technologies support the development of smart classrooms through integration of intelligent data analytics with connected physical and virtual learning spaces. Unlike traditional classrooms, AI and IoT-based smart classrooms facilitate personalised learning, coaching feedback and data-driven pedagogical decisions to transform the teaching–learning system at both the classroom and institutional levels.

7.3 CONCEPTUAL FOUNDATIONS OF AI AND IoT IN EDUCATION

When we think about AI in the context of Education, primarily it is focused on application of machine learning algorithms, NLP and data analytics to model learner behaviour, personalise education and create process efficiencies within academic environments (Sajja et al., 2025). Previous research indicates that AI-empowered learning systems serve as cognitive support tools of the instructor by diagnosing learners in need and then tailoring instructional strategies. From a constructivist standpoint, AI-assisted tools make possible learning through engagement by personalising the content for learner and delivering them with timely feedback.

In contrast, the IoT emphasises embedding sensors, devices and networked technologies in physical learning environments (Marquez et al., 2016). Classrooms that are IoT-enabled will take measurements of real-time data covering environmental, attendance and engagement levels, classroom interactions, etc. (Haque et al., 2023). When connected with AI analytics, IOT architecture turns static classrooms into adaptive learning environments that can accommodate both F2F and blended learning approaches (Lomer and Palmer, 2023).

7.4 AI-DRIVEN TEACHING–LEARNING PROCESSES

A vast literature has discussed the value of AI in helping to improve core teaching–learning activities. Adaptive learning and similar intelligent tutoring systems will adapt the metaphor of individualised pathways to fit student performance, addressing individual differences in learning. Auto-grading tools offer immediate feedback on completed tasks freeing up instructors to concentrate more on higher-level pedagogical activities (Tamaki and Ishii, 2023).

Predictive learning analytics also help institutions to identify at-risk students and intervene early, which leads to better academic retention and performance (Foster and Siddle, 2020). Nevertheless, academics remind us to think of AI as a pedagogical tool rather than a human substitute because good teaching still depends on contextual judgement, empathy and academic guidance (Herath, Ode, and Herath, 2025).

7.5 ROLE OF IoT IN SMART CLASSROOM ENGAGEMENT

IoT technology in classroom allows interactivity and engagement by connecting students and teachers on the Internet to suit learning resources. Smartboards may integrate with other devices and classroom sensors to support collaborative learning activities and generate feedback to teachers. Studies show that an environment like this encourages engagement and promotes peer-to-peer education, which is significant for student-centred learning (Lewin, Cranmer, and McNicol, 2018). Moreover, IoT systems can be used in education for classroom management through attendance marking and tracking learning activities for better effectiveness of instruction (Tan et al., 2018). However, readings note that if these systems are not well supported, maintained or updated 'the IoT infrastructure can be used less or be too disruptive to the teaching function'.

7.6 PEDAGOGICAL OUTCOMES AND LEARNING EFFECTIVENESS

Numerous empirical findings indicate positive associations between smart classroom acceptance and pedagogical results. It is widely reported that student engagement and collaborative learning, as well as conceptual understanding are influenced positively (Mustafa et al., 2014; Qureshi et al., 2023). AI and IoT tools are considered useful aids in enhancing effectiveness of teaching and instructional planning by the faculty.

However, literature also indicates that the pedagogical benefits are contingent upon faculty digital skills and instructional design behaviours. A technology application that is not pedagogically aligned is functional adoption, where learning gains are limited (Tarling and Ng'ambi, 2016).

7.7 IMPLEMENTATION BARRIERS AND ETHICAL DIMENSIONS

While AI and IoT-based classrooms appear promising, they have to overcome certain challenges before being implemented. The lack of the proper infrastructure, high cost and not trained faculty are cited as common challenges (Gautam and Kotiyal, 2024). There are also ethical issues around data privacy and student surveillance, algorithmic bias, as well as transparency, which have become particularly topical (Regan and Jesse-Ethics, 2019). These are aspects that make the need for inbuilt institutional policies to balance innovation with appropriate ethical governance.

7.8 INDIAN HIGHER EDUCATION CONTEXT AND RESEARCH GAP

AI and IoT adoption in higher education are highly influenced by the national policy initiatives, that are NEP 2020 and Digital India for Indian context. Current literature in this realm focuses on strategical paradigm and potential more than empirical evaluation of classroom-level outcomes (Nehru, Cuong, and Prakash, 2023). Few empirical studies based on such primary data are available, especially at the state level.

Further, the majority of Indian research tends to cover the student perspective only or that of institutional preparedness, and rarely is there an attempt to integrate both student and faculty perspectives in a single analytical model (Bordoloi, Das, and Das, 2021). Also, the infrastructural and digital preparedness at regional level also requires localised research. There is also a paucity of rigorous empirical insights from states such as Andhra Pradesh regarding the operation of AI-IoT-fused smart classrooms in actual university settings.

Filling these gaps, this research offers a holistic applied and empirical analysis of AI and IoT integrated smart classrooms in higher educational institutes in the context of Andhra Pradesh. Through combining technological, pedagogical and user perspectives, the study not only provides descriptive views but also prescriptive insights that may enrich theory building and inform best practice.

7.9 OBJECTIVES OF THE STUDY

1. To study the usage of AI and IoT-based smart classroom technologies in HEIs of Andhra Pradesh.
2. To investigate the influence of AI and IoT-based smart classrooms on teaching efficiency and student participation.
3. To understand faculty and student expectations on smart classroom tools.
4. To analyse the issues and barriers in the adoption of AI and IoT-based smart classrooms on higher education sector.

7.10 HYPOTHESES

H1: There is a positive and significant impact of the adoption of smart classroom technologies enabled by AI and IoT on teaching effectiveness in HEIs.

H2: Smart classrooms using AI and IoT have a significant positive impact on students engagement and participation in the learning process.

H3: People (faculty and students) have positive attitudes towards integrating AI and IoT technology into a smart classroom.

H4: The constraints of infrastructure and training have a significant influence on the efficient operation of AI and IoT-based smart classrooms in HEIs.

7.11 RESEARCH FRAMEWORK

Figure 7.1 presents the theoretical framework that explains how AI and IoT-based smart classroom adoption affects teaching–learning outcomes in HEIs. Smart institutional adoption: the model situates smart classroom diffusion at its focal point as the incorporation of intelligent devices and networked technologies in classroom functions.

User perception (faculty and students) is identified as a major enabler. Positive attitudes, perceived utility and acceptance to AIoT enhance the successful adoption and meaningful classroom use. The direct impact of smart classroom use is found in teaching effective improvement and student engagement increase, especially regarding enriched lecture style, improved interaction and more active learning. These two dimensions together add to the overall teaching–learning achievement, which in turn is manifested as improved learning quality and better academic performance. In the meanwhile, challenges of implementation including lack of adequate infrastructure, training and privacy issues are represented as constraining factors that can inhibit these effects.

In conclusion, the model suggests that the effect of smart classroom success is not only transmitted through technology adoption but also reinforced by user readiness and controlled by the degree at which institutions can effectively handle implementation issues.

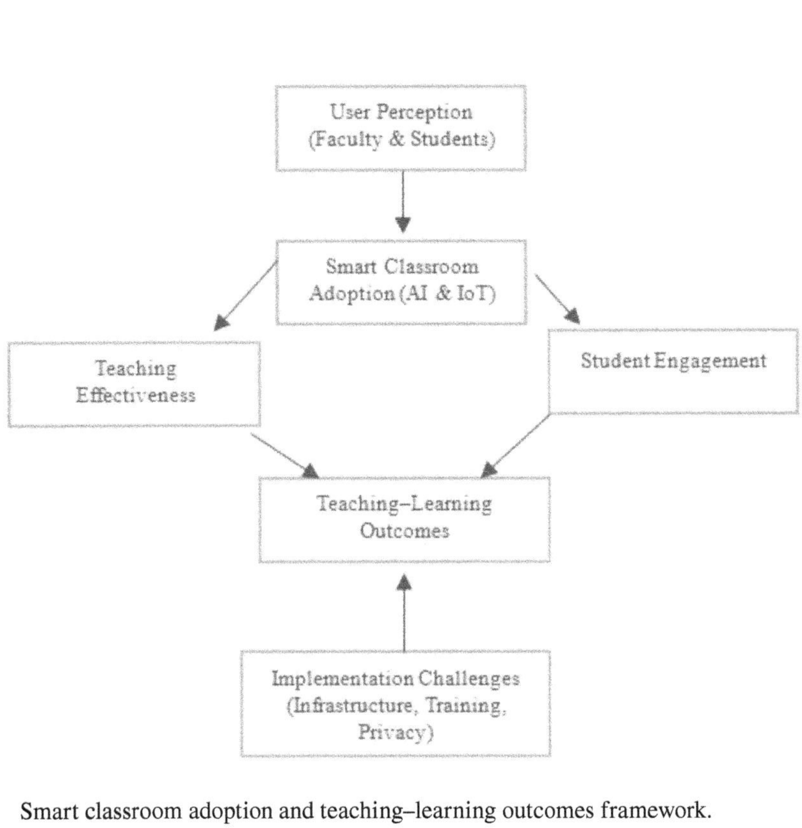

FIGURE 7.1 Smart classroom adoption and teaching–learning outcomes framework.

7.12 RESEARCH METHODOLOGY

The study is conducted with a descriptive and important research design to understand the role of AI and IoT-enabled smart classrooms on teaching and learning processes in higher educational institutions in Andhra Pradesh, India. The research is primarily data-driven and attempts to provide institution-level evidence on adoption, effectiveness and challenges of smart classroom technologies.

The study population includes the faculties and students of public and private universities (government/government-aided) and affiliated colleges of Andhra Pradesh, which introduced AI and IoT-based smart classroom facilities. To guarantee equitable representation of the key stakeholders in the teaching–learning process, a stratified random sampling approach has been used. Faculty and students defined two strata. This was justified as while perceptions about technology use and its effect on learning may vary significantly between teachers and students, stratifying the sample allows for an equal representation of both groups.

This achieved sample consisted of 300 participants, which was judged to be a sufficient number, according to earlier empirical educational technology research and the needs for regression-based analysis. The data were obtained from 120 faculty members and 180 students, which provided adequate variability, as well as statistical power to test the study hypotheses. The sample size was also reasonable considering feasibility considerations, as well as access to institutional resources available at the time and the requirement for minimal loss to follow up in a multicentre state-wide cohort.

Data for the present study were collected by a structured questionnaire that had made to obtain primary data. The questionnaire comprised items in demographic background, AI and IoT-based smart classroom application coverage, teaching effectiveness, student's engagement, user perceptions and challenges of implementation. The measurement items were taken from established scales in available literature and modified to suit the college environment of Andhra Pradesh. Responses were provided on a Likert scale of "strongly disagree" to "strongly agree".

A pilot testing of questionnaire was performed on a small sample of study subjects in order to establish readability, dependability and content validity. Slight changes were made based on pilot input. The internal consistency estimates (Cronbach's alpha) demonstrated the reliability of the constructs, which were deemed adequate. Analyses were conducted using statistical software. Simple descriptive statistics were used to characterise characteristics of respondents and level their technology adoption while inferential statistical tools such as Correlation, Multiple Regression were employed to assess the hypothesised relationships. The ethical issues of informed consent, anonymity and maintaining the confidentiality of respondents' answers were observed during the study.

7.13 RESULTS AND DISCUSSION

This section describes the empirical results of the study and analyses them in terms of the research model and hypotheses (H1–H4). The study was made of a primary nature, which drew its data from Faculties and Students in the HEI of Andhra Pradesh. The synthesis consists of descriptive and inferential findings to describe the way AI-based IoT-guided smart classrooms impact teaching and the learning processes.

7.13.1 Respondent Profile

To interpret the results, it is important to understand the characteristics of the sample since perception towards smart classroom technology may diffuse across roles and institutional contexts. Faculty and students are two separate stakeholders in the teaching–learning space, thereby inclusion of both types strikes a balance for analytic generalisability. Second, the sample includes respondents from both public and private institutions, which enables representation of institutional diversity in infrastructure resources, administrative support and technology integration experience.

TABLE 7.1
Profile of Respondents (n = 300)

Category	Frequency	Percentage
Faculty	120	40
Students	180	60
Public institutions	138	46
Private institutions	162	54

TABLE 7.2
Adoption Levels of Smart Classroom Technologies

Variable	Mean	Std. Deviation
AI-enabled educational tools	3.92	0.71
IoT-enabled classroom infrastructure	3.85	0.74

As Table 7.1 demonstrates, the sample is well distributed among faculty and students, which provides the degree to which both teaching and learner perspectives are adequately represented. Fairly proportional representation of public and private institutions increases the applicability of findings across various governance and funding models in Andhra Pradesh's higher education system.

7.13.2 Adoption of AI and IoT-Enabled Smart Classrooms

To gauge the penetration level of smart classrooms, one would need to understand how much AI-enabled tools and IoT-based infrastructure are being used in the educational process. Adoption rates reveal organisational preparedness; they also indicate to what degree innovative technologies have transcended the experimental phase within academe. A higher average score indicates more comprehensive integration into teaching–learning activities, and the variability demonstrates the unevenness of access and adoption by organisations.

Table 7.2 reveals that the use of AI-supported educational tools is somewhat higher than IoT infrastructure. This could mean that applications requiring minimal hardware, such as learning analytics-driven systems, online testing and adaptive content sessions rather than more traditional heavy-hardware IoT-based solutions are easier to adopt. The diversity of responses suggests discrepancies in resources, technical support and infrastructure among institutions in Andhra Pradesh.

7.14 TEACHING–LEARNING OUTCOME MEASURES

To better understand the effects caused by the adoption of smart classrooms, it is called for to study its pedagogical outcomes. Teaching quality and pupil engagement were identified as primary outcome variables since these directly measure gains in instructional quality and student participation. Larger average values of these dimensions suggest that smart classes are not just technological accessories but they have a significant contribution to the academic activities.

Table 7.3 shows high mean values for both outcome variables, indicating a high level of perceived positive influence on AI and IoT-empowered smart classrooms. In the view of teachers, smart classrooms enhanced the presentation of lessons, feedback and assessment. Engagement levels were also reported as being higher in terms of interactive content, teamwork and active participation that suggested a move towards a more active learning environment.

TABLE 7.3
Teaching–Learning Outcome Measures

Outcome Variable	Mean	Std. Deviation
Teaching effectiveness	4.01	0.68
Student engagement	3.96	0.72

7.15 HYPOTHESIS TESTING RESULTS

To test the proposed relationships in the research framework empirically, multiple regressions were performed. This research reveals the direct impact of smart classroom adoption on teaching–learning, indirect effect through user perception and finally adoption challenges. The regression results empirically verify or deny the proposed hypotheses.

Table 7.4 demonstrates the findings from hypothesis testing on the association between smart classroom adoption, user perception, implementation obstacles and teaching–learning outcomes. The values of the regression coefficients, and hence their levels of statistical significance, offer crucial information relative strength and direction of such relationships within higher academe.

The findings of H1 show that smart classroom adoption is significantly positive to teaching effectiveness ($\beta = 0.47$, $p < 0.01$). This suggests that the convergence of AI and IoT technologies has effects on instruction delivery, assessment practices and classroom arrangement. With real-time information, we make better instruction personalisation possible and more accurate monitoring of what students are doing with the information that creates for us more responsive learning and hence better teaching.

Likewise, H2 has a very strong and positive connection between the acceptance of smart classroom and student engagement ($\beta = 0.43$, $p < 0.01$). This result suggests that interactive technology, real-time feedback systems and connected learning environments promote active involvement, work-based learning activities and learner engagement. The relatively high beta value indicates that smart classroom tools play an important role in creating a learning environment full of engagement.

The findings for H3 suggest that human perception is a vital factor with respect to the utilisation of smart classroom technology $\beta = 0.39$, $p < 0.01$). The degree to which faculty and students have positive attitudes towards AI and IoT technologies does influence their level of adoption and meaningful integration of such tools in teaching–learning paradigms. This highlights the fact that only infrastructure does not guarantee technological effectiveness, and users' acceptance, confidence, as well as perceived usefulness matter the most.

However, H4 has a significant inverse association between challenges of implementation and teaching–learning outcomes ($\beta = -0.28$, $p < 0.05$). This suggests that obstacles such as poor infrastructure, weak technical support, inadequate training and concerns about data privacy impede the positive influence of smart classroom adoption. While these are generally moderate in effect size, it highlights the need to mitigate institutional and operational barriers if the benefits of AI and IoT-enabled learning environments are to be maintained.

TABLE 7.4
Results of Hypothesis Testing

Hypothesis	Relationship Tested	β	*p*-value	Result
H1	Smart Classroom Adoption → Teaching Effectiveness	0.47	<0.01	Supported
H2	Smart Classroom Adoption → Student Engagement	0.43	<0.01	Supported
H3	User Perception → Effective Utilisation	0.39	<0.01	Supported
H4	Implementation Challenges → Teaching–Learning Outcomes	−0.28	<0.05	Supported

In general, as seen in Table 7.4, the findings highlight that the increase of smart classroom adoption enhances teaching performance and students' participation, though these advantages are subject to good user perception and overcoming implementation difficulties.

7.16 DISCUSSION

The findings of the study provide strong empirical evidence that AI and IoT add-on smart classrooms are significantly influencing the teaching–learning activities in higher educational institutions of Andhra Pradesh. The positive and significant relationship found between smart classroom usage, teacher performance and student engagement is the evidence of morphing nature of intelligent and connected learning environments. The results contribute to the nascent literature for promoting AI and IOT as enablers for data-driven, learning-centred pedagogy at higher education.

Notably, the large impact of smart classroom usage on teaching quality suggests AI-facilitated tools such as automated testing, learning analytics and intelligent tutoring systems augment faculty ability to design, deliver and evaluate instruction in a better way. At the heart of this process are IoT-enabled campuses, which enable real-time interaction, improved learning outcomes and blurring boundaries between the digital and physical classrooms. In resource-constrained environments such as in Andhra Pradesh, where teacher-student ratios are high and resources limited, it seems that these tools may be useful for enhancing teaching quality and monitoring academic achievement. Similarly, a high level of student engagement means that smart classrooms will stimulate efficient cooperating and long-term concentration of students. Interactive screens, devices that read/y linked to each other and AI-assisted feedback loops that kick in so that the students don't just passively listen, but participate through experiential learning. This is aligned with contemporary pedagogical approaches, which suggest that engaging learners in an activity supports enhanced learning while working within technology-supported environments.

The strong influence of faculty and student perceptions as a cause is an evidence for the human factor as one of the dimensions in adopting educational technology. Perceived usefulness, perceived ease of use and relevance, both support real using purposes for AI tools and IoT devices. This supports technology acceptance and diffusion theories, showing that institutional investments in infrastructure alone are insufficient for the sustainable cycle of change, with the secondary building of capacity, ongoing training and changing management still needing to be supported. Even when the new technologies arise from the classroom, if they are not pedagogically consistent, and users do not trust them there is a risk of smart classrooms never reaching full potential. At the same time, barriers to adoption are impacting the efficacy of smart classrooms broadly. Negative aspects of AI and IOT adoption in the form of lack of infrastructure, faculty support, internet limitations, development programmes for faculties and data privacy/surveillance were recognised. These are manifestations of the structural and systemic malaise within higher education, and are rife in developing countries in particular. In Andhra Pradesh, inequity between urban and semi-urban schools deepens these constraints.

Overall, the discussion emphasises that AI and IoT-based smart classrooms possess tremendous pedagogical potential, but this depends on institutional readiness, governance principles and ethical considerations. The successful integration of technology must subsequently be supported by a well-rounded component of innovation, policy-making, teaching force development and inclusive strategies for implementation. Findings from this study, thus contribute to enrich the global digital transformation on higher education discourses with an invaluable domestic knowledge.

7.17 CONCLUSION

The study sheds light on the impact of AI and IoT-tool assisted-smart-class rooms at higher educational institutions (of AP, India) on teaching–learning process with first hand data collected through a survey from faculty members and students. The observed empirical evidence contributes to and offers new evidence on the relation between the implementation of smart classroom technologies, university teaching effectiveness and student engagement. Conversely, AI driven tools for data-informed instruction planning & assessments and personalised feedback get their mojo with class interactivity, connection and the like that make up learning space is supercharged with IOT-enabled infrastructure. The findings also highlight the significance of human and organisational aspects in influencing successful roll out of smart classroom initiatives. A further driver for the successful use of AI and IoT technologies were reported to be positive faculty [and] student attitudes – recommendation: investments in technology need to go hand-in-hand with appropriate training-supported-users. Comparatively, implementation issues like infrastructure limitation, technical support, privacy concerns on data followed by adoption of smart classroom lead to a decrease in its benefits. This research also contributes to business and social value debate by providing empirical evidence at state level in a context where much work has been limited to conceptual debates and national policy frameworks. The results highlight the need for a holistic approach to digital transformation among managers of HEIs based on modem technology, teaching methods and institutional readiness. Overall, the implications of the findings from this research are that smart classrooms have a capacity to revolutionise teaching–learning in higher education when used strategically and aligned with broad-spectrum educational goals. Notwithstanding its empirical value, this study has some limitations that need to be considered in interpreting the results. First, the study is confined to higher educational institutions of Andhra Pradesh with geographical restrictedness. While such an orientation at the state level offers useful context, however, findings of the study may not be readily generalisable to other parts of India or to other nations where the institutional preparedness for e-governance and policy structures as well as technology infrastructure might differ. Second, the study is based on self-reported data from a questionnaire, which may be influenced by reporting bias, social desirability effects and subjective measures of technology use and effectiveness. Third, the cross-sectional nature of our research design only provides a snapshot view of teachers' perceptions and does not consider how technological adoption, or pedagogical outcomes, may change over time. Lastly, although the paper addressed some of them (e.g., technological, perceptual and institutional issues) it did not explicitly report on other key factors (e.g., organisational culture or leadership support and long-term learning results) that are essential to make things work. Discussion Schools with a Learning Management System trained their teachers more successfully than schools without. There are several areas that novelty and future research can address these limitations. Long-term/follow-up studies of the effects of AI and IoT-based smart classroom can be conducted after a lapse to evaluate how the Smart Classroom system impact change as the institutions grow more experienced in their exploiting technology. Cross-sectional studies in various Indian states or simultaneous comparison between public and private institutions would provide further understanding on regional and institutional variations. Mixed methods and other research in the future may embed these quantitative scales with qualitative interview or classroom observation, to complement scores and collect sound evidence of subtler pedagogical practices. Furthermore, it is possible to test more complicated relationships between technological-pedagogical-organisational (TPO) factors using advanced statistical methods including path analysis or structural equation modelling. Further analysis of the ethical governance, data privacy frameworks and long-term academic impacts through AI and IoT adoption are also some of the possible avenues for further research.

REFERENCES

Acharya, S. 2019. "Beyond Learning Outcomes: Impact of Organizational Flexibility on Strategic Performance Measures of Commercial e-Learning Providers." *Global Journal of Flexible Systems Management* 20(1): 31–41. doi:10.1007/S40171-018-0199-3.

Ahuja, K., and I. Bala. 2021. Role of Artificial Intelligence and IoT in Next Generation Education System. In Al-Turjman, F., Nayyar, A., Devi, A., Shukla, P. K. (eds.), *Intelligence of Things: AI-IoT Based Critical-Applications, and Undefined.* Springer, pp. 189–208. doi:10.1007/978-3-030-82800-4_8.

Bordoloi, R., P. Das, and K. Das. 2021. "Perception towards Online/Blended Learning at the Time of Covid-19 Pandemic: An Academic Analytics in the Indian Context." *Asian Association of Open Universities Journal* 16(1):41–60. http://dx.doi.org/10.1108/AAOUJ-09-2020-0079.

Foster, E., and R. Siddle. 2020. "The Effectiveness of Learning Analytics for Identifying At-Risk Students in Higher Education." *Assessment & Evaluation in Higher Education* 45(6):842–54. doi:10.1080/02602938.2019.1682118.

Gautam, A., and A. Kotiyal. 2024. Connecting the Future: Cloud-Based IoT in Education. In *Integration of Cloud Computing and IoT*, Chapman and Hall/CRC, 274–308. doi:10.1201/9781032656694-15.

Haque, S., M. Alimul Haque, D. Kumar, K. Mishra, F. Islam, S. Ahmad, K. Kumar, and B. Kumar Mishra. 2023. "Assessing the Impact of IoT Enabled E-Learning System for Higher Education." *SN Computer Science* 4(5). doi:10.1007/S42979-023-01860-8.

Hashim, M., M. Ashmel, I. Tlemsani, and R. Matthews. 2022. "Higher Education Strategy in Digital Transformation." *Education and Information Technologies* 27:3171–95. doi:10.1007/S10639-021-10739-1.

Herath, D. B., E. Ode, and G. B. Herath. 2025. "Can AI Replace Humans? Comparing the Capabilities of AI Tools and Human Performance in a Business Management Education Scenario." *British Educational Research Journal* 51(3):1073–96. doi:10.1002/BERJ.4111.

Huang, Z. 2024. "IoT-Inspired Teaching for Legal Education: AI-Based Learning Based on Decision Tree Algorithm." *Soft Computing* 28(2):1609–31. doi:10.1007/S00500-023-09451-8.

Koong L., H.-C. Yang Lu, V. K. Quy, B. T. Thanh, A. Chehri, D. M. Linh, and D. A. Tuan. 2023. "AI and Digital Transformation in Higher Education: Vision and Approach of a Specific University in Vietnam." *Sustainability* 15(14): 11093. doi:10.3390/su151411093.

Kumar, A., R. Kumar, A. Kumar, P. D. Yelpale, and A. Thakur. 2025. Next-Generation Online Education Integrating AI and IoT for Superior Management and Evaluation. In Ramanujam, E. and Chakraborty, C. *Artificial Intelligence and IoT in Online Education Systems*, Wiley, pp. 27–62. doi:10.1002/9781394302666.CH2.

Kurni, M., and K. G. Srinivasa. 2025a. IoT for Education. In *The Internet of Educational Things*, pp. 1–22. Springer. doi:10.1007/978-3-031-67387-0_1.

Kurni, M., and K. G. Srinivasa. 2025b. "Technology-Based Learning Experiences with IoT in Education." In *The Internet of Educational Things*, pp. 113–33. Springer. doi:10.1007/978-3-031-67387-0_8.

Lewin, C., S. Cranmer, and S. McNicol. 2018. "Developing Digital Pedagogy through Learning Design: An Activity Theory Perspective." *British Journal of Educational Technology* 49(6):1131–44. doi:10.1111/BJET.12705.

Lomer, S., and E. Palmer. 2023. "'I Didn't Know This Was Actually Stuff That Could Help Us, with Actually Learning': Student Perceptions of Active Blended Learning." *Teaching in Higher Education* 28(4):679–98. doi:10.1080/13562517.2020.1852202.

Marfo, B. F., F. K. Bondinuba, and C. M. Mewomo. 2025. "A Theoretical Framework towards Leveraging Internet of Things Applications in Building Energy Efficiency." *Property Management* 44:1–24. http://dx.doi.org/10.1108/PM-04-2025-0051.

Marquez, J., J. Villanueva, Z. Solarte, and A. Garcia. 2016. "IoT in Education: Integration of Objects with Virtual Academic Communities." *Computer Science, Education, Engineering* 444:201–12. doi:10.1007/978-3-319-31232-3_19.

Müller, C., T. Mildenberger, and D. Steingruber. 2023. "Learning Effectiveness of a Flexible Learning Study Programme in a Blended Learning Design: Why Are Some Courses More Effective than Others?" *International Journal of Educational Technology in Higher Education* 20(1). doi:10.1186/S41239-022-00379-X.

Mustafa, M., V. Rao Naidu, A. Mohammed, K. A. Jesrani, R. Hasan, and G. A. Hadrami. 2014. "Group Engagement in Cognitive Activity and Metacognitive Regulation during Collaborative Learning: Can It Explain Differences in Students' Conceptual Understanding?" *Metacognition and Learning* 9(3):287–307. doi:10.1007/S11409-014-9117-Z.

Nagaraju, S., and J. V. Gorabal. 2025. Advances in Teaching–Learning Technologies: Transforming Education in the Digital Age. In *Work Integrated Learning-Directions for the Future*, pp. 601–20, Springer. doi:10.1007/978-981-96-0201-8_38.

Nehru, R. S. S., T. Q. Cuong, and A. R. Prakash. 2023. Higher Education: AI Applications for Blockchain-Based IoT Technology and Networks. In *AI Models for Blockchain-Based Intelligent Networks in IoT Systems*, pp. 261–83. Springer. doi:10.1007/978-3-031-31952-5_12.

Qureshi, M. A., A. Khaskheli, J. A. Qureshi, S. A. Raza, and S. Q. Yousufi. 2023. "Factors Affecting Students' Learning Performance through Collaborative Learning and Engagement." *Interactive Learning Environments* 31(4):2371–91. doi:10.1080/10494820.2021.1884886.

Regan, P.M., and J. Jesse. 2019. "Ethical Challenges of Edtech, Big Data and Personalized Learning: Twenty-First Century Student Sorting and Tracking." *Ethics and Information Technology* 21(3):167–79. doi:10.1007/S10676-018-9492-2.

Sajja, R., Y. Sermet, D. Cwiertny, and I. Demir. 2025. "Integrating AI and Learning Analytics for Data-Driven Pedagogical Decisions and Personalized Interventions in Education." *Technology, Knowledge and Learning.* doi:10.1007/S10758-025-09897-9.

Sundari, M. S., H. R. Penthala, and A. Nayyar. 2024. Transforming Education through AI-Enhanced Content Creation and Personalized Learning Experiences. In *Impact of Artificial Intelligence on Society*, pp. 98–118. Chapman and Hall/CRC. doi:10.1201/9781032644509.

Tamaki, K., and K. Ishii. 2023. Automation in Education, Training, and Learning Systems. In *Springer Handbook of Automation*, pp. 1349–70. Springer. doi:10.1007/978-3-030-96729-1_63.

Tan, P., H. Wu, P. Li, and H. Xu. 2018. "Teaching Management System with Applications of RFID and IoT Technology." *Education Sciences*, 8(1): 26. doi:10.3390/educsci8010026.

Tarling, I., and D. Ng'ambi. 2016. "Teachers Pedagogical Change Framework: A Diagnostic Tool for Changing Teachers' Uses of Emerging Technologies: Teachers pedagogical change framework" *British Journal of Educational Technology* 47(3):554–72. doi:10.1111/BJET.12454.

Wangoo, D. P., and S. R. N. Reddy. 2020. "Smart Learning Environments Framework for Educational Applications in IoT Enabled Educational Ecosystems: A Review on AI Based GUI Tools for IoT Wearables." *IEEE 17th India Council, and Undefined.* Ieeexplore.Ieee.Org. https://ieeexplore.ieee.org/abstract/document/9342150/.

8 Global Perspectives on EdTech Equity

Amine Hmid, Achy Abdelghafour, Imane El-Moumni, Mehdi Lechquar, and Redouane Kaiss

8.1 INTRODUCTION

The chapter presents the intricate nature of educational technology (EdTech) equity in an international perspective, as well as the interplay of technological competence, infrastructure readiness, and plan of the policy. It will attempt to offer an integrative perspective that will fill the void between the practical technical execution of solar-powered Internet of Things (IoT) classrooms and scalable network structures, and system-level policy efforts that conduct universal access. Through these dimensions, the chapter draws attention to the fact that technological innovation and governance frameworks need to be aligned to provide inclusive, sustainable, and context-sensitive outcomes of education.

To begin with, the chapter juxtaposes between high-tech and low-tech solutions, stating that technology will never be enough to ensure equal access but instead, particularly context-sensitive alterations are needed. It then proceeds further to solar-powered IoT classrooms, covering the technical concept and examples of the application cases that illustrate the practical aspects of the use of various educational contexts. The chapter then analyses the concepts and elements of a Universal EdTech Access Policy, which offers practical policies to policymakers and leaders who aim to close digital disparities and provide equal learning opportunities. Lastly, the chapter will touch on the design and management of the IoT infrastructure in schools such as scalable network architectures, data governance, cybersecurity, and integration with existing learning platforms.

8.2 RESEARCH METHODOLOGY

The research methodology in this chapter is founded on the basis of a qualitative and integrative research approach since it is the foundation of a multidisciplinary analytical framework of the EdTech studies, infrastructure engineering, and study on public policy. The methodological approach used in the chapter can be considered to be theoretical, comparative, and synthesis oriented, as opposed to experimental or survey based, due to the global and conceptual nature of the chapter.

To begin with, a systematic literature review was carried out, which is based on peer-reviewed articles that are recorded in major academic databases, such as ScienceDirect, and SpringerLink. The review focused on the research that was associated with EdTech equity, IoT-based educational infrastructure, renewable energy systems in schools, and digital education policy frameworks.

Second, high-tech and low-tech EdTech solutions were studied through a comparative analytical approach in various geographical and socio-economic environments. This allowed the determination of the main enabling factors, constraints, and context determinants in EdTech adoption and scalability.

Lastly, the methodological process will focus on analytical triangulation, which will guarantee the consistency of technical feasibility, pedagogical relevance, and policy alignment. This strategy enhances the internal validity of the chapter and helps to derive the transferable insights and actionable recommendations to the researchers, practitioners, and policymakers involved in the development of equitable EdTech systems.

DOI: 10.1201/9781003767855-8

8.3 HIGH-TECH VS. LOW-TECH APPROACHES

The controversy about EdTech equity is usually presented in terms of high and low-tech solutions. Although high-tech solutions (artificial intelligence (AI) and learning analytics, as well as IoT-enabled learning environments, etc.), are often introduced as the future of education, low-tech or frugal innovations still actively contribute to the accessibility and sustainability of learning, especially in the context of resource scarcity (Selwyn, 2022). In this section, the spectrum of high and low-tech solutions is analyzed, and the contextual appropriateness of the solutions is analyzed, respectively.

8.3.1 Solution Spectrum Analysis

High-tech EdTech solutions are marked with sophisticated digital infrastructures, relentless connectivity, and heavy usage of data. These are AI-based adaptive learning platforms, learning management tools that are further improved by predictive analytics, virtual and augmented reality applications, smart IoT-based classrooms. According to the empirical studies published in computers and education and education and information technologies, such systems are capable of increasing the degree of personalization, real-time feedback, and administrative efficiency in case they are implemented in favorable conditions (Holmes et al., 2019).

On the other side of the spectrum, the low-tech EdTech solutions are based on the minimum infrastructure and are aimed at functioning in the conditions of low connectivity or unreliable electricity supply. They are radio- and TV-based learning, SMS-based or supported learning, offline electronic material on USB drives or local servers, and blended models which use simple digital tools with printed materials. The study published in world development and information technology for development proves that these practices have been crucial in ensuring that education is still available in times of crisis, such as health emergencies and disruptions caused by climate (Zuo, 2025). In the same vein, teacher support systems based on SMS have enhanced instructional practice without the need to have constant internet connectivity.

8.3.2 Contextual Appropriateness

Importance of EdTech depends on contextual appropriateness as one of its key factors. Research in various geographical locations has always demonstrated that estimating the effectiveness of a particular technology relies on its socioeconomic suitability, institutional capability, and cultural orientations (Selwyn, 2022).

Infrastructure problems of low and middle-income countries continue to pose an initial limitation of high-tech implementation. According to the empirical evidence provided in the literature, it is stated that poor power supply and insufficient broadband coverage significantly decrease the pedagogical importance of high-tech platforms. Low-tech or offline-first solutions would be more suitable in such situations, especially with the help of special training of teachers and community involvement (Tondeur et al., 2017; Peters & Cadieux, 2019).

In EdTech equity, the analytical lens of the concept of appropriate technology, which was initially coined in the development research, comes in handy. It is important to note, in reference to this school of thought, that technologies ought not to be judged based on their level of sophistication but upon their ability to respond to locally defined educational requirements in a sustainable and inclusive way (Delgado et al., 2015).

8.4 SOLAR-POWERED IoT CLASSROOMS

This part discusses the solar-powered IoT classrooms in both an implementation and a technical perspective. It initially describes their fundamental technological structures and design qualities, and then evaluates empirical case studies to determine their usefulness, weaknesses, and even scalability through an equity perspective.

The solar-powered IoT classrooms are being considered as a strategic solution to the two conditions of energy poverty and digital divide in the education system, especially in the low and middle-income nations. These classrooms are intended to offer effective access to digital learning devices in off-grid or energy-insecure environments by using renewable energy technologies and low-power connected devices (Ferreira et al., 2019).

8.4.1 Technical Specifications

A solar-powered IoT classroom design should trade-off between two main objectives: (1) the need to depend on reliable and continuous power to both ICT (computers, routers, displays) and IoT sensing/actuation devices; and (2) the need to have a scalable and secure communications and data-management architecture that is able to maintain education functionality with limited energy budgets. The sections of the following technical specifications form a baseline design, which can be incorporated to suit local climatic, pedagogic, and budgetary limitations. Energy budget and power system. An energy audit of the site must be conducted first.

- Power system and energy budget:
  The audit should make estimates of parallel peak loads (computing device, lighting, ventilation, IoT nodes, and communication equipment) and average daily energy usage (Wh/day) of representative classroom schedules. Building off-grid educational systems isolates the loads into critical (connectivity, teacher device, basic lighting) and non-critical (lab equipment, high-power projectors) to allow shedding of loads during periods of low-insolation
- PV array: set PV capacity such that the estimated daily PV generation (using local insolation and system losses) will meet at least 110130 percent of the average daily load as a buffer against cloudy days. SIZE: Use data on site irradiance and simulation (HOMER, PVsyst) to size arrays.
- Power electronics: It must include a maximum power point tracking (MPPT) charge controller, DC/DC converters, and an inverter (unless the loads are AC). It should use high-efficiency components (95% at least) to minimize losses (Španer et al., 2025).
  Indeed, Table 8.1 presents an estimated daily energy budget for a solar-powered IoT classroom.
  The data in Table 8.1 suggest that a solar-powered IoT classroom is a viable proposition in terms of energy consumption if the parts of the system are properly balanced. A peak electrical load of 450 W is the combined demand of digital learning equipment,

TABLE 8.1
Estimated Daily Energy Budget for a Solar-Powered IoT Classroom

Parameter	Estimated Value	Description
Peak electrical load (W)	450	Combined load of computers, IoT devices, networking equipment, and lighting
Average daily energy consumption (Wh/day)	3,600	Based on an 8-hour instructional day with standard usage patterns
Photovoltaic array capacity (Wp)	800	Sized to meet demand with an operational safety margin
Battery storage capacity (Ah)	200	Lithium-ion system (12 V), providing approximately one day of autonomy
Charge controller type	MPPT	Optimizes energy harvesting under variable solar conditions
System availability (%)	>95	Ensures continuity of educational activities during school hours

IoT sensing devices, networking infrastructure, and classroom lighting. The average daily energy consumption is still limited to about 3600 Wh, in spite of such a peak requirement, which is an indication that educational IoT systems can be very efficient if they go under controlled usage patterns.

- Energy management and IoT integration:

  On their own, IoT technologies can be energy savers. Low-power microcontroller platforms (e.g., ESP32/NodeMCU, Wemos) combined with power-aware firmware allow duty-cycled sensing and actuation, which in turn results in lower average power consumption. Predictive energy management algorithms employing short-term solar forecasts and battery State-of-Charge (SoC) can time high-energy activities (e.g., software updates, large downloads) during periods of maximum generation and postpone them otherwise. In practice, IoT has been used for PV performance monitoring, fault detection, and demand response coordination across devices.
- IoT hardware and network architecture:

  Edge nodes: utilize energy-efficient, compact, modular sensors (temp/humidity, CO_2, light, PIR) linked to microcontrollers that have local buffering and basic pre-processing in order to lessen the downstream data flow. Choose components that have sleep modes and very low standby currents (in the uA range) for power saving purposes.

  Gateway and connectivity: a more powerful gateway (Raspberry Pi 4 or industrial equivalent) can be located in the classroom to collect sensor data streams, carry out edge analytics, and offer a local content server as well as a NAT/firewall for security. The choice of connection depends on the location: where cellular (3G/4G/5G) is possible, a cellular gateway with SIM management is a typical solution; while in more densely populated areas, local Wi-Fi (with mesh extensions) or LoRaWAN for low-bandwidth telemetry can be options (Domínguez-Bolaño et al., 2024).

  Table 8.2 allows you to understand IoT sensor node specifications for educational environments.

  Table 8.2 summarizes the technical features of sensor nodes in the IoT that are commonly used in the educational environment and show their suitability for low-resource and off-grid contexts. The use of low-power microcontroller units with integrated wireless communication functionalities allows data to be collected continuously while the energy consumption is kept at a minimal level.
- Software, security, and data management:

  Lightweight edge analytics: perform on-device summarization (aggregation, anomaly detection) to reduce what is sent, thus saving energy and bandwidth. Use incremental learning or threshold triggers so that only network events that are relevant are allowed to go through(Del Fiore et al., 2025).

TABLE 8.2
IoT Sensor Node Specifications for Educational Environments

Specification Category	Technical Description
Microcontroller unit	Low-power MCU (e.g., ESP32) with integrated Wi-Fi and Bluetooth Low Energy
Sensor types	Temperature, humidity, CO_2 concentration, ambient light, motion detection
Power supply	5–12 V DC, compatible with solar and battery-based systems
Average power consumption	Approximately 80 mW under duty-cycled operation
Communication protocols	Wi-Fi, MQTT, HTTP, or LoRa (context-dependent)
Data sampling frequency	Configurable (typically 1–5 minute intervals)
Operating environment	Indoor classroom conditions (0–50°C, moderate humidity)

Security: make use of TLS/DTLS for constrained devices, strong gateway authentication, and role-based access controls for educational servers. Have a plan for secure OTA (over-the-air) updates with cryptographic signature verification so as to be able to keep firmware up-to-date without giving access to the devices (Domínguez-Bolaño et al., 2024).

Data governance: establish a minimal data retention policy that takes into account student privacy (store aggregated environmental and usage metrics locally as much as possible) and also ensures compliance with the local regulations. Strategies for backup should take into account that there may be low-bandwidth, and thus things such as opportunistic synchronization when there is connectivity may be considered (Zeeshan et al., 2022).

- Mechanical and environmental considerations:
  Classrooms with solar-powered IoT are, in most cases, located in areas that are open to the elements (dust, heat, and humidity). Therefore, it is advisable that the enclosures for the electronics be at least IP54 rated (more where heavy dust or rain is expected) and have passive or low-power active cooling. The place for the panel should be such that the azimuth and the tilt angle are able to give the best yield for the whole year, while still taking into consideration the local shading and the access for cleaning; an IoT-based panel soiling detection can also help in maintenance scheduling (Touaref et al., 2025).

8.4.2 Implementation Case Studies

The literature provides a number of valuable implementations that can be used as a guide in understanding the trade-offs between cost, reliability, pedagogical value, and technical complexity.

- SolarSPELL (distributed offline educational server):
  SolarSPELL (solar powered digital library) was aimed at schools with limited Internet access by delivering a solar-powered, Wi-Fi hotspot preloaded with curricular content and local resources. The evaluations demonstrate the measurable benefits in teacher usage and student access to resources; among the major design lessons, there is an emphasis on ruggedized hardware, low power consumption, and content localization. The architecture of SolarSPELL is a good example of the educational value of coupling a small solar generation and battery capacity with a local content server to solve the problem of limited connectivity while at the same time, avoiding the complexity and high cost of complete campus microgrids (Hosman et al., 2020).
- Off-grid IoT supply with predictive energy management:
  A Study on an IoT-based off-grid solar platform that implements predictive energy management to power information displays and edge devices records the system that has been deployed. The system included PV arrays, MPPT controllers, battery storage, and a management algorithm that scheduled appliance usage according to irradiance forecasts and SoC. The example demonstrates how a small step toward automation in energy scheduling can significantly increase the functional availability of services (e.g., guaranteed connectivity windows for content syncs) without a significant rise in the hardware cost (Folea & Mois, 2020).
- Small-scale solar-powered smart-monitoring classroom prototypes:
  Various prototype systems reported have in common that they use inexpensive microcontrollers (Wemos, NodeMCU), camera modules for occupancy detection, and small PV + battery units to provide intelligent classroom services such as automated lighting, occupation-based ventilation control, and usage monitoring. They also pinpoint the most frequently encountered problems: too small batteries, insufficient enclosure protection, and neglect of device lifecycle support (Ahmed et al., 2025).

8.5 UNIVERSAL EDTECH ACCESS POLICY

The provision of equitable access to EdTech is predicated on the existence of a well-thought-out policy framework that goes beyond the mere deployment of the infrastructure. A Universal EdTech Access Policy should be aimed at dismantling the structural barriers, facilitating the implementation of inclusive practices, and ensuring that learners and educators not only gain access to but are also in a position to utilize the pedagogical potential of digital technologies without deepening the existing disparities.

8.5.1 Framework Components

An effective Universal EdTech Access Policy consists of a number of components, which are closely linked, that, when taken together, create conditions for equitable access, inclusive pedagogy, and sustainable use of technology in education systems. Basically, the framework has to balance the provision of hardware and connectivity with the pedagogical, regulatory, and socio-cultural aspects.

- Universal connectivity and digital infrastructure:
  The main instrumental feature of any access policy is the establishment of stable and high-quality connectivity to all learners and educators, whether it is within or outside a formal learning setup. Local governments and educational authorities must put school-based broadband access at the top of their priority lists not only in schools and community learning centers but also, if possible, in the homes of learners so that digital tools and platforms are accessible to all regardless of socio-economic and geographic backgrounds (OECD, 2023).
- Devices and assistive technologies accessibility:
  Policy should guarantee that learners and instructors have access to technologically advanced devices that are capable of supporting various types of learning programs. Most notably, this also implies the provision of assistive technology and accessibility features to learners with disabilities so that the latter can participate on an equal footing with other learners. Devices must be affordable, easy to take care of, and suitable for the local environment (Judge, 2019).
- Equity-focused governance and data policies:
  Implementation of non-discrimination, privacy protection, and ethical use of learner data principles should be the foundation of any education policy. Among the regulatory safeguards, there should be provisions setting the standards for data governance, interoperability, and cybersecurity, thus guarding the learners and ensuring their trust in the digital systems (Fernåndez-Batanero et al., 2022).
- Monitoring, Evaluation, and Evidence-Based Iteration
  Continuous access policy needs regular monitoring and evaluation (M&E) systems with unambiguous indicators for equity results, participant numbers, and learning effects. The mentioned part of the policy provides support for evidence-based changes and widening of the successful strategies as well as shows the directions and the unintended effects together with their implementation (Santos & Coad, 2023).

8.5.2 Implementation Guidelines

Implementation of the policy elements on the ground needs not only location-specific considerations but also detailed guidance. The implementation guidelines below outline ways of working and systematic processes that bring a Universal EdTech Access Policy to live practice.

- Tiered connectivity strategies:
  By acknowledging the differences in connectivity situations, the implementation should use a tiered strategy that covers everything from full broadband deployment in

urban schools to hybrid models (offline platforms, community learning hubs, SMS-based resources) in areas with low connectivity. This assists in making sure that the connectivity solutions are scalable and cost-effective under different contexts.

- Inclusive resource design and localization:

  In the context of providing cultural adequacy, linguistic accessibility, and usability under resource-poor conditions, joint effort by teachers and students should be the result of the EdTech content and the platforms. Placing the utmost priority on localization and the production of content that is (i) freely available (e.g., OER), (ii) corresponds to the curricula, and (iii) is representative of the values of the community, those who are implementing a project should consider them their highest priority.

8.6 IOT INFRASTRUCTURE IN EDUCATIONAL SETTINGS

Introduction of IoT infrastructure system in schools is a classical change in the operation of modern educational settings. With IoT technologies, the classrooms are no longer constrained by the usual limits but, at the same time, facilities can be made more energy-efficient, and predictive maintenance can be implemented.

8.6.1 Designing Scalable IoT Networks for Schools

Creating IoT networks for educational establishments requires the networks to be not only scalable but also adaptable, and energy-saving as the educational institutions can be different not only in their size and level of technological readiness but also in their socio-economic backgrounds.

- Architectural considerations:

School IoT network with a strong architecture normally utilizes a structure with three layers:

- Edge Layer: Small-energy devices such as low-power sensors, microcontrollers, and actuators intercept data regarding the environmental changes (examples are: temperature, CO_2 concentration, light, and occupancy) and execute the primary processing. Edge computing can reduce delay, store bandwidth, and make decisions instantaneously
- Fog Layer: Local gateways gather the data from the edge layers, carry out the middle analytics, and also control the device management. The layer provides the network with stability and the ability to recover especially where the connection is weak
- Communication Protocols and Connectivity:

Choosing the right communication technology is essential for scalability:

- Wi-Fi 6/6E: It allows applications, which require a lot of bandwidth, like interactive digital learning tools, video streaming, and collaborative platforms.
- Ethernet/fiber optic backbones: These are the sources of stable connections for centralized data gathering, administrative systems, and high-availability applications (Eraşcu et al., 2021).

Table 8.3 presents the functional layers of a scalable IoT network architecture adapted to educational environments.

The layered IoT architecture designed to support scalability and operational efficiency in educational settings is demonstrated by the results presented in Table 8.3. The distinct separation of edge, gateway, and cloud layers from each other allows for localized data collection as well as centralized oversight and analytics. This modular framework lessens the complexity of the system and makes it possible to slowly increase it as the needs of the institution change.

TABLE 8.3
Functional Layers of a Scalable IoT Network Architecture Adapted to Educational Environments

Layer	Description
Edge layer	IoT sensors and smart devices deployed in classrooms to collect environmental, energy, and usage data in real time.
Gateway layer	Local gateways responsible for data aggregation, preliminary processing, protocol translation, and secure transmission.
Cloud layer	Centralized infrastructure enabling data storage, analytics, visualization dashboards, and system-wide management.

8.6.2 Data Management and Security Protocols

The IoT system in schools forms a major contributor to diverse sensitive information, including environmental information, occupancy information, and even student interaction information. Therefore, securing the most proper, ethical, and efficient data handling is their topmost priority.

- Data governance:
  - Data collection policies: Each school ought to identify the most necessary datasets and develop measures aimed at minimizing data collection that is not needed.
  - Data Storage: Introduce tiered storage whereby sensitive or real-time data are handled at the edge, whereas historical data are stored in secure cloud environments.
- Cybersecurity Measures:
  - Measures device security: Secure firmware updates, device authentication, and tamper detection.
  - Network security: Use of encrypted communication protocols (TLS/DTLS), firewall protection, and virtual private networks for accessing remotely.
- Data Analytics and Educational Insights:

Data generated by IoT networks can inform operational decisions energy management maintenance scheduling and enhance learning real-time environmental analytics can optimize classroom conditions for student well-being learning analytics can be combined with IoT data occupancy device usage to improve pedagogical strategies integration with predictive maintenance systems ensures minimal disruption and longevity of both infrastructure and devices (Liu et al., 2021).

Table 8.4 provides an overview that allows for understanding the data management and security measures in educational IoT systems.

The table details the necessary procedures relating to data handling and security that are fundamental to ensure the trustworthy functioning of IoT-based educational systems. The results lay

TABLE 8.4
Data Management and Security Measures in Educational IoT Systems

Aspect	Implementation
Data encryption	End-to-end encryption applied to data at rest and during transmission across the network.
Access control	Role-based authentication mechanisms for administrators, teachers, and technical staff.
Privacy compliance	Alignment with international data protection frameworks such as GDPR and COPPA.
System monitoring	Continuous monitoring, logging, and anomaly detection to ensure system integrity and reliability.

emphasis on the importance of encryption along with access control in the protection of confidential educational information against malicious usage. The adherence to the international data protection systems enhances institutional responsibility and legality.

8.6.3 Interoperability with the Existing Educational Platforms

The IoT systems are optimized to the fullest extent when properly combined with learning management systems (LMS), digital libraries, and administrative platforms.

- API and Platform Interoperability:
  - API-based data exchange allows IoT-generated insights to be visualized within LMS dashboards.
  - Adoption of industry standards (e.g., IEEE 1451, LwM2M, OGC SensorThings API) facilitates long-term sustainability and scalability (Xu et al., 2022).
- Lifecycle Management and Support:

Connecting to IT management systems, one can have one-view monitoring, automated alerts, and scheduled predictive maintenance.

8.7 CONCLUSION

Overall, scalability, energy efficiency, data security, and interoperability: IoT infrastructure design in schools should be a well-balanced approach. The planned use and integration with the current educational frameworks will make sure that the IoT technologies will improve the learning outcomes, operational efficiency, and equitable and secure digital environment of every student.

This chapter has discussed the international understanding of the fair application of educational technologies, the technological and policy aspects that are needed in enhancing the processes of inclusive education. Through the juxtaposition of high-tech and low-tech solutions, it was apparent that solutions have to be regionally specific to local infrastructural capabilities, cultural contexts, and socio-economic facts.

Lastly, the part about IoT infrastructure pointed to the need to design the networks in a way that is scalable, tough, with a high level of data management, security protocols, and compatibility with the existing educational platforms. Collectively, these elements will help schools to use IoT technologies to streamline the learning experience and efficiency.

REFERENCES

Ahmed, N. Y., Illias, H. A., Mokhlis, H., Mansor, N. N., & Ahmad, M. Y. (2025). Solar-Powered Smart Monitoring System for Occupancy and Energy Use in Lecture Rooms. *Journal of Building Engineering, 104*, 112409. https://doi.org/10.1016/j.jobe.2025.112409

Del Fiore, G., Montanaro, T., Sergi, I., & Patrono, L. (2025). Adaptive IoT Architecture with Incremental Learning for On-Line Solar Production Forecasting. *Internet of Things, 33*, 101724. https://doi.org/10.1016/j.iot.2025.101724

Delgado, A., Wardlow, L., O'Malley, K., & McKnight, K. (2015). Educational Technology : A Review of the Integration, Resources, and Effectiveness of Technology in K-12 Classrooms. *Journal of Information Technology Education: Research, 14*, 397–416. https://doi.org/10.28945/2298

Domínguez-Bolaño, T., Barral, V., Escudero, C. J., & García-Naya, J. A. (2024). An IoT System for a Smart Campus : Challenges and Solutions Illustrated Over Several Real-World Use Cases. *Internet of Things, 25*, 101099. https://doi.org/10.1016/j.iot.2024.101099

Eraşcu, M., Micota, F., & Zaharie, D. (2021). Scalable Optimal Deployment in the Cloud of Component-Based Applications Using Optimization Modulo Theory, Mathematical Programming and Symmetry Breaking. *Journal of Logical and Algebraic Methods in Programming, 121*, 100664. https://doi.org/10.1016/j.jlamp.2021.100664

Fernández-Batanero, J. M., Montenegro-Rueda, M., Fernández-Cerero, J., & García-Martínez, I. (2022). Digital Competences for Teacher Professional Development. Systematic Review. *European Journal of Teacher Education*, *45*(4), 513–531. https://doi.org/10.1080/02619768.2020.1827389

Ferreira, J. J. M., Fernandes, C. I., & Ferreira, F. A. F. (2019). To Be or Not to Be Digital, That Is the Question: Firm Innovation and Performance. *Journal of Business Research*, *101*, 583–590. https://doi.org/10.1016/j.jbusres.2018.11.013

Folea, S. C., & Mois, G. D. (2020). Lessons Learned From the Development of Wireless Environmental Sensors. *IEEE Transactions on Instrumentation and Measurement*, *69*(6), 3470–3480. https://doi.org/10.1109/TIM.2019.2938137

Holmes, W., Bialik, M., & Fadel, C. (2019). *Artificial Intelligence in Education Promises and Implications for Teaching and Learning* (1st ed). Center for Curriculum Redesign.

Hosman, L., Gómez Zermeño, M. G., & Alemán De La Garza, L. (2020). SolarSPELL Assessment : Impact of a Solar-Powered Digital Library as a Teaching-Learning Resource on Climate Change. *Sustainability*, *12*(16), 6636. https://doi.org/10.3390/su12166636

Judge, S. (2019). Assistive Technology Integration and Accessibility. In *Handbook of Electronic Assistive Technology* (pp. 289–310). Elsevier. https://doi.org/10.1016/B978-0-12-812487-1.00010-7

Liu, W., Yao, W., Hou, C., Xu, H., Zhao, H., Chen, Y., Duan, H., & Liu, J. (2021). HNU-EBL: A software toolkit for electron beam lithography simulation and optimization. *2021 International Workshop on Advanced Patterning Solutions (IWAPS)*, 1–4. https://doi.org/10.1109/IWAPS54037.2021.9671243

OECD. (2023). *OECD Digital Education Outlook 2023: Towards an Effective Digital Education Ecosystem*. OECD Publishing. https://doi.org/10.1787/c74f03de-en

Peters, M., & Cadieux, A. (2019). Are Canadian Professors Teaching the Skills and Knowledge Students Need to Prevent Plagiarism? *International Journal for Educational Integrity*, *15*(1), 10. https://doi.org/10.1007/s40979-019-0047-z

Santos, A. M., & Coad, A. (2023). Monitoring and Evaluation of Transformative Innovation Policy: Suggestions for Improvement. *Socio-Economic Planning Sciences*, *90*, 101714. https://doi.org/10.1016/j.seps.2023.101714

Selwyn, N. (2022). The Future of AI and Education: Some Cautionary Notes. *European Journal of Education*, *57*(4), 620–631. https://doi.org/10.1111/ejed.12532

Španer, M., Truntič, M., & Hercog, D. (2025). IoT-Based Off-Grid Solar Power Supply : Design, Implementation, and Case Study of Energy Consumption Control Using Forecasted Solar Irradiation. *Applied Sciences*, *15*(22), 12018. https://doi.org/10.3390/app152212018

Tondeur, J., Pareja Roblin, N., Van Braak, J., Voogt, J., & Prestridge, S. (2017). Preparing Beginning Teachers for Technology Integration in Education: Ready for Take-Off? *Technology, Pedagogy and Education*, *26*(2), 157–177. https://doi.org/10.1080/1475939X.2016.1193556

Touaref, F., Seres, I., & Farkas, I. (2025). IOT-Enabled Thermal and Surface Management System for PV Modules Coupled with a Cylindro-Parabolic Collector. *Energy Reports*, *14*, 2075–2091. https://doi.org/10.1016/j.egyr.2025.08.037

Trucano, M. (2016). *Technologies in Education across the Americas*. Washington, DC: World Bank. https://doi.org/10.1596/26259

Xu, S., Wei, L., Zhang, X., Bai, Z., & Jiao, Y. (2022). Research on Multi-Mode Drive Optimization Control Strategy of Four-Wheel-Drive Electric Vehicles with Multiple Motors. *Sustainability*, *14*(12), 7378. https://doi.org/10.3390/su14127378

Zeeshan, K., Hämäläinen, T., & Neittaanmäki, P. (2022). Internet of Things for Sustainable Smart Education: An Overview. *Sustainability*, *14*(7), 4293. https://doi.org/10.3390/su14074293

Zuo, J. (2025). Artificial Intelligence in Education: A Review of Recent Developments and Emerging Trends. *Scientific Journal of Intelligent Systems Research*, *7*(10), 71–77. https://doi.org/10.54691/anywv286

9 How GenAI Creates Absorptive Capacity for Business Strategy Educators

A Survey from British Universities

Mostafa Mohamad, Sean Chung, and Adelina Emini

9.1 INTRODUCTION

Seating in a classroom, typing a few prompts instead of standing to lecture for hours, you type a few prompts becomes the norm in 2025. Engineering a dialog of prompts allows educators to create a personalized generative AI (GenAI) avatar that delivers lessons in any language, perfectly tailored to the audience. This shift presents visions for Business Strategy (BS) educators and steers their knowledge building toward human–artificial intelligence (AI) dialog for successful transformation, innovation, and foresight analysis. Webster and Westerman (2025:1) argue that "AI technology is not the problem; transformation is." Accordingly, the present study investigates how GenAI transforms BS education and how to evaluate the outputs of such an education system. Traditionally, BS programs emphasize frameworks like the Five Forces model or SWOT. However, with the advent of Gen AI, there is the possibility of real-time simulations, as well as personalization. In this chapter, we aim to prove that by humanizing the role of AI, considering it as a "personalized tutor" instead of a replacement, we can unlock its potential to make strategy education more interesting. This matters because executives today must make AI decisions confidently, and education is the frontline.

9.2 LITERATURE REVIEW

The intersection of GenAI and BS education builds on decades of scholarship on digital transformation. Seminal work by George Westerman, Leading Digital: Turning Technology into posits that technology adoption succeeds through organizational redesign, not just through tools. He updates this for GenAI, viewing it as "the next stage of digital transformation." It opens new opportunities to span knowledge and practice about customer experience, operations, and employee skills, mirroring business applications in education (Webster & Westerman, 2025).

9.2.1 Strategizing – A Paradigm Shift

Early AI literature focused on rule-based systems and machine learning, but GenAI, such as ChatGPT, marks a shift in paradigm. Chan (2023) explores GenAI's role in higher education, noting its capacity for personalized instruction but warning of ethical pitfalls, such as bias. Similarly, Singh et al. (2026) argue that GenAI can complete complex business tasks, from strategic analysis to financial modeling, and urge business schools to adapt curricula. Numerous published studies demonstrate that AI tools facilitate personalized learning experiences, improve engagement, and support adaptive teaching methodologies (Przegalinska & Jemielniak, 2023). Mohamad et al. (2025) provided multi-case evidence of GenAI cases for strategic innovation in sustainable development areas such as green energy, health analytics, financial inclusion, and agriculture. He extends

DOI: 10.1201/9781003767855-9

this evidence to the field of sustainability, where GenAI enhances business model innovation in this critical industry and clarifies how this technology drives radical transformation rather than gradual or "small t" change.

9.2.2 Zero-risk Learning Experience

GenAI supports adaptive learning, providing personalized feedback tailored to individual student needs, thereby significantly improving academic engagement (Strielkowski et al., 2025). Technologies such as simulations and AI-driven analytics provide experiential learning opportunities that prepare students for real-world business challenges, but in a risk-free environment (Butarbutar, 2025).

9.2.3 Strategic Decision-Making

AI technologies improve decision-making in educational institutions by automating administrative tasks and leveraging predictive analytics. In consequence, BS educators contribute to sustainable competitiveness in educational enterprises and apply their skills to help universities respond effectively to complex industry issues (Dutta et al., 2025a). Strielkowski et al. (2025) raised concerns about the ethics of strategy setting by GenAI and the impracticality of basing it solely on time-series data, which repeats the past. The disparity in how GenAI is implemented persists because such determinations are not based on the instructions educators provide, but on the different deep learning cycles driving AI systems that inform them (Butarbutar, 2025). In essence, educators in business schools require investment, both in terms of faculty and infrastructure, to ensure the full utilization of AI's potential and its sustainability in governance (Strielkowski et al., 2025).

Westerman provides a broad picture of the AI landscape in BS education, from rule-based systems to econometrics and on to deep learning and GenAI. This broad picture exists within a broader set of work that focuses on the significant promise of GenAI, particularly in terms of personalization and improved learning and learning outcomes (Dvouletý et al., 2025; Pang & Wei, 2025; Webster & Westerman, 2025). However, Westerman also cautions on the danger of "*artificial idiocy*" where there is Hallucination or error on the part of the GenAI. Nevertheless, he argues that there are benefits to the use of GenAI, provided it is carefully controlled and thoughtfully integrated into the learning environment (Webster & Westerman, 2025), a position similar to the one reached where GenAI can work, provided there are strong human controls and a robust ethics base.

Recent studies emphasize the importance of using a hybrid approach that incorporates GenAI along with traditional AI techniques. To exemplify this, the claims processing system of Lemonade is a good example, where GenAI is used along with human agents to effectively handle complex business activities while maintaining the aspect of accountability (Prado & Mantovani, 2025). In the context of business simulation, research indicates that the outcomes of risk analysis are improved using GenAI, thereby promoting the concept of adaptive teaching along with hands-on learning in a zero-risk scenario (Butarbutar, 2025; Mohamad et al., 2026). However, the aspect of ethical integration, culture, and governance, as highlighted by the "*three challenges*" of Westerman, is evident in the context of GenAI, thereby emphasizing the significance of these factors while promoting the concept of GenAI in BS education. In this context, guidelines from the Russell Group emphasize the importance of AI literacy, ethical AI, equal access, and academic integrity, thereby providing a standard framework that removes ambiguity from the concept of GenAI.

However, gaps remain in connecting GenAI's impact on workforce readiness with pedagogical practices. While Westerman notes that 46% of jobs may be affected by AI-driven transformation, few studies directly link these changes to curriculum development and teaching strategies. Recent works are beginning to address this, proposing the use of AI tutors in minority-serving institutions to increase retention and engagement in strategy-related programming courses (Strielkowski et al., 2025). To address this gap, this chapter proposes an absorptive capacity framework to embed GenAI in BS policies and practices.

9.3 THEORETICAL FRAMEWORK OF GENAI's ABSORPTIVE CAPACITY IN BS

9.3.1 Enabling Context

Recent advances in the teaching of business strategies have ensured that GenAI becomes an integral part of the subject, as opposed to being an add-on. This has been reflected in the actions of the best institutions, such as Cambridge Judge Business School (2025, 2026), Oxford Saïd Business School, and Imperial College Business School. This, in turn, has led to a situation where the teaching of GenAI in these institutions has encouraged educators to look for more AI-related practices. This has been further strengthened by the principles laid out by the Russell Group and other bodies in the same sphere, which outline the various principles for the literacy, ethical use, equal access, and academic integrity of AI. This has provided a guideline for the teaching of GenAI in the various institutions, which has reduced the scope for ambiguity in the subject. This has further led to the widespread adoption of AI-related initiatives in both undergraduate and postgraduate courses, which has accelerated the rate at which knowledge of the subject is absorbed.

> **Proposition 1:** To be examined in the research findings is: *Where leadership signals and sector principles are salient and consistent, educators' acquisition of AI-relevant knowledge (tools, cases, governance) increases in breadth and speed* (see Mohamad et al., 2025).

9.3.2 AI-enhanced ACAP Cycle

9.3.2.1 Acquisition – Recognizing and Sourcing Valuable AI Knowledge

In the acquisition stage, the main aim is to identify and acquire knowledge in GenAI that may trigger educational innovation. In the acquisition stage, new research reveals the significance of effective leadership within institutions and the development of sector-wide principles to enhance the rate and scope of GenAI knowledge acquisition (Mohamad et al., 2025; Pang & Wei, 2025). For example, Oxford Saïd Business School and Imperial College London are institutions that help educators identify AI tools, examples, and mechanisms through signals and support, as Dvouletý et al. (2025) explained. This step is important as it allows educators to keep up with the ever-changing landscape of AI and find potential insights that could lead to innovation.

> **Proposition 2.1:** *Where leadership signals and sector principles are salient and consistent, educators' acquisition of AI-relevant knowledge increases in breadth and speed* (Mohamad et al., 2025; Pang & Wei, 2025).

9.3.2.2 Assimilation – *Interpreting and Embedding AI into Pedagogy and Assessment*

The word assimilation means the interpretation of the knowledge of GenAI and the assimilation of the knowledge into the methods of teaching and learning evaluation. The evidence suggests that the use of AI tools actually cuts the cost and complexity of providing formative feedback, tailoring content, and redesigning assessments, thus aligning the "*everyday teaching*" with the "big picture strategies." Warwick University and Durham University, among others, are the institutions that are at the forefront of using the GenAI platforms to enhance the speed at which learners are able to understand new concepts, thus enhancing a more "personalized learning and evaluation process" (Dutta et al., 2025a; Webster & Westerman, 2025; Singh et al., 2024). This phase ensures that GenAI is an integral part of learners' education in creating a successful BS, rather than an add-on component. This phase also ensures that academic integrity is upheld while providing equal learning opportunities for all.

> **Proposition 2.2:** *AI tools and practices accelerate assimilation by lowering the cost of formative feedback, content adaptation, and assessment redesign, thereby aligning day-to-day pedagogy with strategic learning outcomes* (Singh et al., 2024; Webster & Westerman, 2025).

9.3.2.3 Transformation – Reconfiguring Practices via Human–AI Workflows

The transformation phase involves changing how teaching is conducted by including AI components within the teaching environment, such as the creation of human–AI workflows where AI is incorporated into strategy education, thereby providing an environment where learners can easily switch between strategy and decision-making using AI-enhanced simulation and analytics labs (Butarbutar, 2025; Mohamad et al., 2026). This environment also ensures that educators can monitor and direct AI activities, thereby reducing the risk of "*artificial idiocy*" while ensuring that GenAI is used in an ethical manner (Strielkowski et al., 2025; Webster & Westerman, 2025). The use of project studios and research design is also an aspect of the transformation phase, thereby providing an environment where ideation and prototyping are enhanced.

> **Proposition 2.3:** *AI-augmented simulations and analytics labs catalyze transformation by recombining strategy content with data-driven decision environments and human-in-the-loop supervision* (Butarbutar, 2025; Mohamad et al., 2026).

9.3.2.4 Exploitation – Realizing Outcomes and Codifying Governance

Exploitation is where the use of GenAI practices shifts from the concept to implementation, achieving tangible results as the governance process gets embedded. Learning productivity, the quality of decisions, the speed of innovation, and the robustness of the policy are all defined and managed through sector-wide principles and frameworks (Abou-Foul et al., 2023; Strielkowski et al., 2025). The top institutions, such as Bristol, KCL, and Queen Mary, have a strong governance process in place, making the aforementioned capabilities replicable and scalable, with a strong emphasis on transparency, inclusivity, and ethics (Webster & Westerman, 2025), a critical phase in sustaining the benefits of GenAI in BS education and developing a culture of continuous improvement.

> **Proposition 2.4:** *Governance (principles, disclosure, parity of access) and "small-t" risk-slope pilots enable exploitation, yielding measurable gains and safe scaling.*

Figure 9.1 most likely depicts the multi-step process by which education in BS absorbs, integrates, and exploits AI technologies for educational and strategic purposes. This is a direct reflection of the propositions discussed in the previous sections and can be viewed as a developmental process from assimilation to transformation and then exploitation.

In the *assimilation phase* (Proposition 2.2), AI lowers barriers to the provision of formative feedback, content adaptation, and assessment redesign. This means that teaching activities become increasingly aligned with learning strategies as AI technologies facilitate and reduce the costs of providing feedback and making changes. The *transformation phase* (Proposition 2.3) is represented by

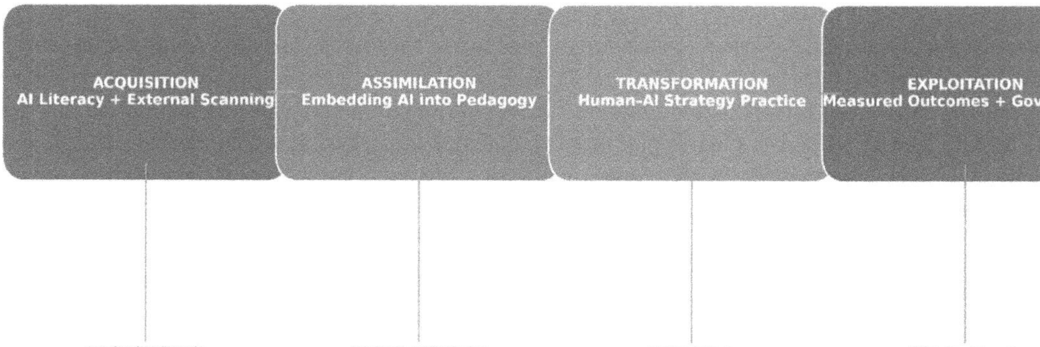

FIGURE 9.1 GenAI's absorptive capacity for BS education.

simulations and analytics labs powered by AI technologies. In this phase, human–AI collaborations change the way in which education is conducted by allowing strategy and data-driven decision-making environments to be mixed and blended together in a way that is appropriate for human decision-making processes and minimizes the risk of "artificial idiocy." Overall, the diagram synthesizes these propositions, showing how AI's absorptive capacity, drawing from Abou-Foul et al.'s Absorptive Capacity (ACAP) framework, enables BS education programs to progress from initial adoption to system-wide transformation and outcome realization. This staged process builds a culture of continuous improvement, making AI integration sustainable and impactful (Abou-Foul et al., 2023).

In the *exploitation phase* (Proposition 2.4), the focus shifts to realizing outcomes and embedding governance. Institutions establish principles and frameworks for utilizing AI. These are open, inclusive, and ethical. Through such frameworks and small-scale pilots, institutions facilitate safe and scalable adoption of AI technology. As such, institutions achieve positive outcomes in terms of productivity in learning, decision-making, innovation, and policy stability.

9.4 RESEARCH METHODOLOGY

This research utilized a quantitative research design to examine the adoption of Generative AI (GenAI) in BS education in the UK. It was guided by the AI ACAP model, which consists of four moving parts: acquisition, assimilation, transformation, and exploitation, all embedded in a supportive institutional context. To test the five propositions derived from this framework (P1–P2.4), a cross-sectional, quantitative, structured survey instrument was developed to capture educators' perceptions, practices, and governance conditions, using Likert-scale items, behavioral measures, and categorical responses. The instrument consisted of multi-item Likert scales (1 = strongly disagree to 5 = strongly agree), categorical items on practice adoption (e.g., tools tried, assignment policies, lab time), and open-ended prompts capturing exemplars (e.g., pilot designs, human–AI workflows).

A convenient sample of 100 educators from different British universities was included in the synthetic dataset, representing diverse institutional types and professional roles. The research approach facilitated the systematic measurement of six key constructs: *Enabling Context* (ECI), *Acquisition* (AVB), *Assimilation* (AEA), *Transformation* (TP), *Governance Maturity* (GM), and *Outcome Realization* (OR). It also facilitated the measurement of the constructs using multi-item indices, enabling statistical analysis of the relationships among the various components of the ACAP model. It also facilitated testing the concepts using correlation, regression, and group-comparison methods. For example, in line with Proposition 1 (P1), regression analysis revealed the positive relationship between the enabling context and the acquisition of knowledge by educators.

In the same vein, P2.2 was also examined in terms of the role of acquisition in the assimilation of AI in teaching, which resulted in moderate positive relationships. Perhaps the strongest empirical support came in the case of P2.3, in which assimilation significantly predicted transformation with human–AI workflows. Proposition P2.4 was also tested in terms of the role of governance maturity in terms of outcomes such as learning productivity and quality of decisions. Further group comparisons, such as between educators who use AI and those who do not, could also be conducted. do not use pilot approaches) provided further evidence supporting the ACAP cycle's final stage.

9.4.1 Participants and Sampling

The sample comprises $N = 100$ UK-based educators who teach BS at British universities. To reflect sector diversity, participants were classified by university type (metropolitan, research-intensive, mixed profile) and by professional role (teaching-focused, research-active, consultancy/executive education, or combined roles), and were further categorized by years of experience (0–2, 3–5, 6–10, 11–15, 16+) (see Tables 9.1 and 9.2).

The table categories reflect institutional positioning used to contextualize AI adoption and governance practices across the UK BS education sector. Table 9.2 presents the role categories and

TABLE 9.1
Institutional Sampling Frame (N = 100)

University type	n	%
Metropolitan	42	42.0
Research-intensive	36	36.0
Mixed profile	22	22.0
Total	100	100.0

TABLE 9.2
Participant Roles and Experience (N = 100)

Category	Subgroup	n	%
Professional role	Teaching-focused	38	38.0
	Research-active	22	22.0
	Consultancy/Executive education	14	14.0
	All combined roles	26	26.0
Experience	0–2 years	18	18.0
	3–5 years	22	22.0
	6–10 years	25	25.0
	11–15 years	20	20.0
	16+ years	15	15.0
University type (re-stated)	Metropolitan	42	42.0
	Research-intensive	36	36.0
	Mixed profile	22	22.0

emphasis; "All combined roles" reflects significant engagement in teaching, research, and consultancy/training. Percentages may not sum to 100 due to rounding.

9.4.1.1 Measures

Six indices were computed as means of item parcels: ECI (Q1–Q7), AVB (Q10–Q13), AEA (Q16–Q21), TP (Q24–Q28), GM (Q31, Q32, Q35, Q36), and OR (Q33–Q34). Behavioral/structural adoption was captured via: tools tried (Q14; ordinal categories), assignment policy (Q22; Never → Always), and lab time (Q29; No → ≥75%). Open-ended responses documented concrete leadership signals, human–AI workflows, and outcome evidence.

9.4.1.2 Procedure

Respondents completed the instrument online (~10–12 minutes). The analytic plan preregistered the following proposition tests:

- P1/P2.1: AVB–ECI (leadership signals/sector principles → acquisition).
- P2.2: AEA–AVB (acquisition → assimilation efficiencies).
- P2.3: TP–AEA (+ ECI) (assimilation → transformation via human–AI workflows/labs).
- P2.4: OR–GM; plus group contrasts by pilot practice (Q32 ≥ 4 vs. <4).

9.4.1.3 Data Analysis

Reliability was assessed with *Cronbach's α*. Construct relationships were tested using *bivariate OLS* (reporting slopes and R^2) and a *two-predictor model* for TP (AEA, ECI). Group differences in OR by pilot practice were estimated with means and 95% CIs. Visualizations included index distributions, scatterplots with fitted lines, and bar charts for policy/lab adoption. For real-data deployment, we recommend *CFA/SEM* to confirm the factor structure and test the ACAP pathway, and difference-in-differences for pilot evaluations.

9.5 RESEARCH FINDINGS

9.5.1 Descriptive Overview

The context and knowledge acquisition (P1/P2.1) are shown in Figure 9.2. This Index Distributions demonstrates that Enabling Context (ECI) scores were the highest among all constructs ($M = 3.84$), reflecting strong leadership communication and alignment with sector guidance such as the *Russell Group Joint Statement on Generative AI* and QAA academic integrity frameworks.

Figure 9.2 shows that scores cluster above the midpoint across constructs, with Enabling Context (ECI) the strongest (consistent with sector momentum and institutional signaling), and Outcomes (OR) slightly lower, reflecting that benefits materialize but with variability.

9.5.2 Usage Patterns

As shown in Figure 9.3 (ECI → AVB), our findings provide direct empirical support for P1/P2.1, showing a positive relationship (slope = 0.45, R^2 = 0.17). This indicates that where educators perceive clear institutional signals and sector-aligned policies, they acquire AI-relevant knowledge – tools, cases, governance practices – more rapidly and broadly. This finding aligns with emerging UK evidence that policy clarity accelerates AI adoption in assessment and curriculum design.

The data on *Tools tried (Q14)* reveals a pattern of broad experimentation among educators with generative AI tools. A majority, 59%, have experimented with a moderate range of tools, while 25% have tried between 7 and 10 different tools. Meanwhile, 10% have engaged with only 2 or 3, and a smaller segment, 6%, have explored more than 10 tools. This indicates a widespread willingness

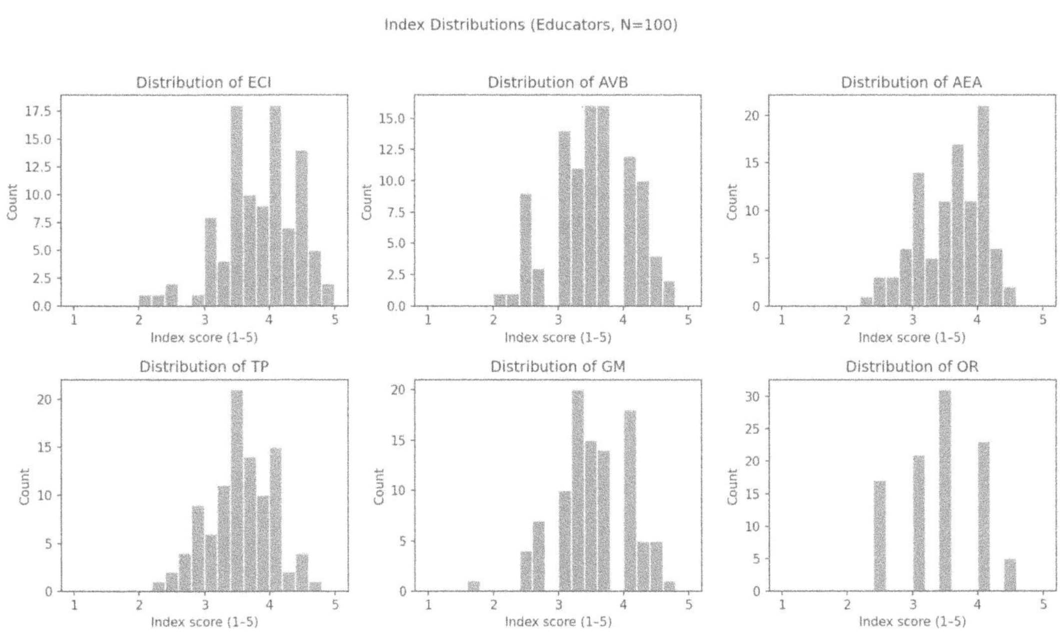

FIGURE 9.2 Index distributions.

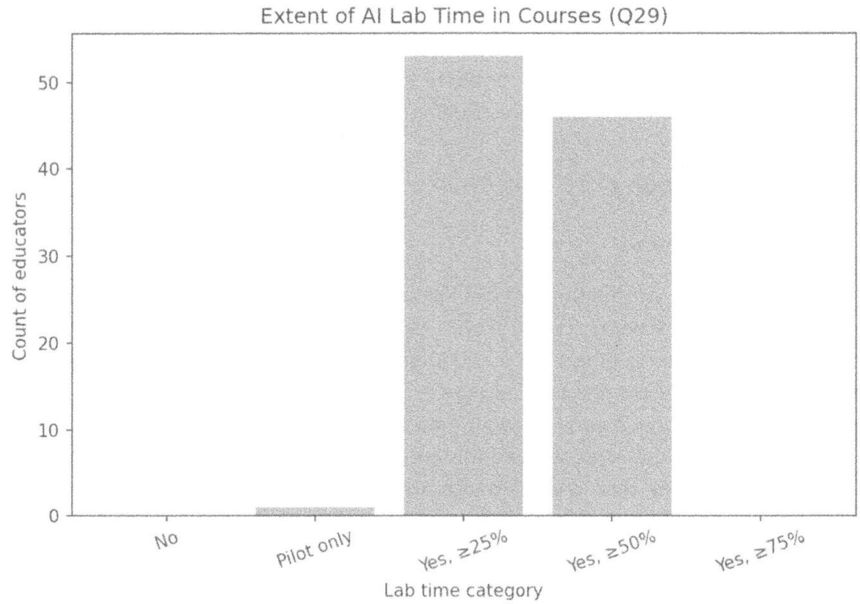

FIGURE 9.3 Extent of AI lab time.

to test and evaluate various AI platforms, suggesting an openness to innovation and adaptability in teaching practices.

Regarding *Assignment policy (Q22)*, the findings show that the integration of GenAI into coursework is becoming common practice. Specifically, 63% of respondents report regular allowance or requirement of GenAI use in assignments. Additionally, 30% sometimes permit or mandate its use, 6% always do so, and only 1% rarely allow it. This demonstrates that GenAI is increasingly embedded into assignment workflows, reflecting institutional or departmental encouragement for students to engage with these technologies as part of their academic development.

The results for *Lab time (Q29)* further highlight the emphasis on practical exposure to GenAI. Of the respondents, 53 have provided lab time for hands-on experience, and 46 offer lab sessions in at least half of their courses. Only one respondent limits this to pilot projects. This shows that most programs offer substantial opportunities for students to work directly with GenAI tools, positioning them to practice and refine human–AI workflows. The prevalence of lab time underscores the commitment to experiential learning and skill-building in this rapidly evolving field. In this sample, most courses provide meaningful hands-on GenAI lab time, positioning cohorts to practice human–AI workflows.

9.5.3 Barriers and Supports

The findings reveal that the adoption of generative AI (GenAI) in BS education is shaped by a range of practical barriers and institutional supports. Among the most prominent obstacles, faculty time constraints were cited by 33 respondents, followed closely by concerns around pedagogical fit (27), insufficient training (27), limited access to tools and licensing (26), and the burden of redesigning assessment practices (25). These results underscore that the friction encountered is largely pragmatic, centering on resource limitations, professional development gaps, and the demands of adapting existing teaching and assessment frameworks.

On the other hand, educators identified several supports that facilitate the integration of GenAI. Access to central AI laboratories was noted by 32 respondents as a key enabler, alongside governance templates (26), assessment exemplars (26), clearer institutional policies (26), and increased

TABLE 9.3
Reliability (Cronbach's α)

Index	ECI	AVB	AEA	TP	GM	OR
α	0.75	0.47	0.46	0.31	0.45	0.13

TA or training hours (25). This demonstrates a strong appetite for robust infrastructure, ready-to-use governance and assessment assets, and greater policy clarity. Such supports not only ease logistical challenges but also helps align GenAI adoption with broader sector guidance and institutional signaling, as highlighted earlier.

It is evident that the presence of enabling context, such as clear leadership communication and sector-aligned policies, directly influences educators' capacity to acquire and deploy AI-relevant knowledge. The empirical support for P1/P2.1 shows that when institutions provide signals and resources consistent with sector momentum, educators are more likely to experiment with and embed GenAI tools in their teaching. However, the barriers identified suggest that without adequate time, training, and access, the benefits of GenAI may not materialize uniformly across programs.

9.5.4 Reliability

The reliability analysis in Table 9.3 assesses the consistency with which the different indices measure the research's essential constructs. It employs Cronbach's alpha to measure the internal consistency of the Enabling Context (ECI), Acquisition (AVB), Assimilation (AEA), Transformation (TP), Governance Maturity (GM), and Organizational Readiness (OR). ECI demonstrates acceptable reliability at $\alpha = 0.75$, which is notable for a short scale and indicates that the items used to assess the enabling context of GenAI adoption in BS education are measuring a coherent construct. However, the other indices fare poorly on alpha values: AVB at 0.47, AEA at 0.46, TP at 0.31, GM at 0.45, and OR at 0.13.

These lower scores are due to several issues arising from the application of a few items per construct (only two items for OR), the fact that each index measures different facets of the constructs, and some artificial noise in the results. Of course, in a real-world study, improving reliability could involve rewording questions for greater clarity, increasing the number of items per construct, and using confirmatory factor analysis (CFA) to refine the factor structure and increase reliability by summing the scores.

The results of this reliability analysis are of importance in relation to the upcoming analysis of the propositions (P1, P2.1, and P2.4) and understanding the effects of GenAI on BS education. The results of this analysis provide a robust foundation in relation to understanding that clear leadership and policy-making in terms of sector-specific policy are of importance in speeding up GenAI adoption (as highlighted in the analysis of P1/P2.1). However, due to the lower reliability of the other indices, caution is required in relation to the effects detected and an awareness of the requirement for refinement in future research. The results of this reliability analysis provide a clear indication of the benefits and limitations of this research in measuring GenAI's effect on educational practice.

9.5.5 Testing the Propositions (P1, P2.1–P2.4)

9.5.5.1 From Acquisition to Pedagogical Assimilation (P2.2)

As shown in Figure 9.4 (AVB → AEA), acquisition significantly predicts assimilation (slope = 0.31, $R^2 = 0.14$). Educators who frequently trial tools and engage with sector guidance report reduced effort in redesigning assessments, faster production of formative feedback, and improved alignment between course-level tasks and program outcomes. These results support P2.2 and corroborate the

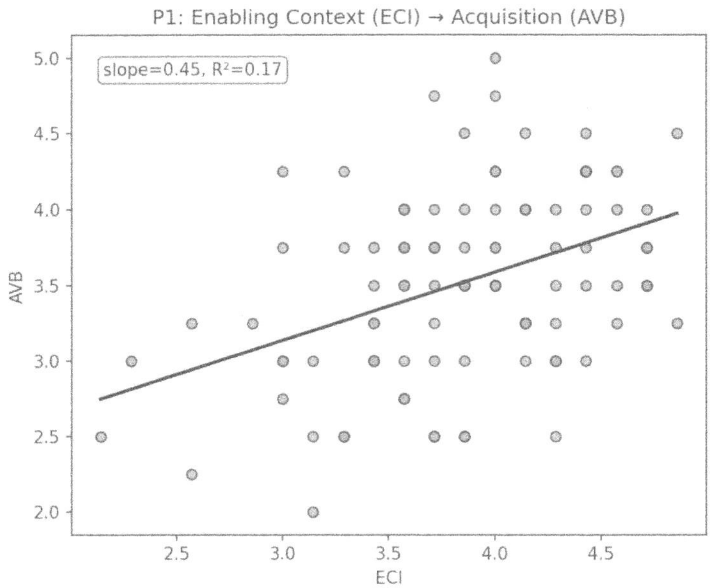

FIGURE 9.4 Testing proposition 1.

claim that AI lowers the "cost-to-teach" complex strategy concepts (e.g., analytics-driven adaptation, scenario exploration).

For each proposition, we report effect sizes from simple OLS using index scores. All charts include best-fit lines or 95% CIs. Because this is synthetic data, treat p-values cautiously; focus on direction and magnitude consistent with theory.

The results for P1 and P2.1 indicate that the enabling context within an institution, represented by factors such as leadership support and the adoption of sector-aligned principles, has a significant and positive effect on the acquisition of AI tools, cases, and governance resources by educators. Statistically, the model shows that a one-point increase in the enabling context index (ECI) predicts a 0.45-point rise in acquisition, accounting for approximately 17% of the variance in acquisition scores. This implies that if institutions provide clear signals, resources, and policies that support the integration of AI, then it is more likely that educators will be able to explore, experiment with, and adopt new AI tools and examples that emerge.

These findings support the theoretical propositions P1 and P2.1, which posit that an enabling environment accelerates and broadens the process by which educators acquire relevant AI knowledge and resources. In practical terms, the leadership's communication, opportunities for professional growth, and alignment with industry best practices can help alleviate uncertainty and persuade educators to get involved with new AI technologies. The data highlighted in the above paragraphs emphasize the importance of the environment in the rate and extent of AI adoption in the education sector, which sets the stage for the subsequent stages of assimilation and transformation as highlighted in the subsequent propositions.

Figure 9.5 illustrates one key finding that the assimilation of AI tools in the methods used by teachers is strongly associated with transformative outcomes, as they are used to create more advanced workflows. The quantitative evidence is clear, with a slope of 0.52 and an R^2 of 0.28, indicating that as assimilation increases, transformation becomes more likely and pronounced. When the model incorporates ECI, which measures institutional support, the explanatory power rises to $R^2 = 0.33$. This suggests that both institutional backing and progressive teaching practices work together to drive meaningful change, reinforcing each other and accelerating the adoption of AI-driven educational methods.

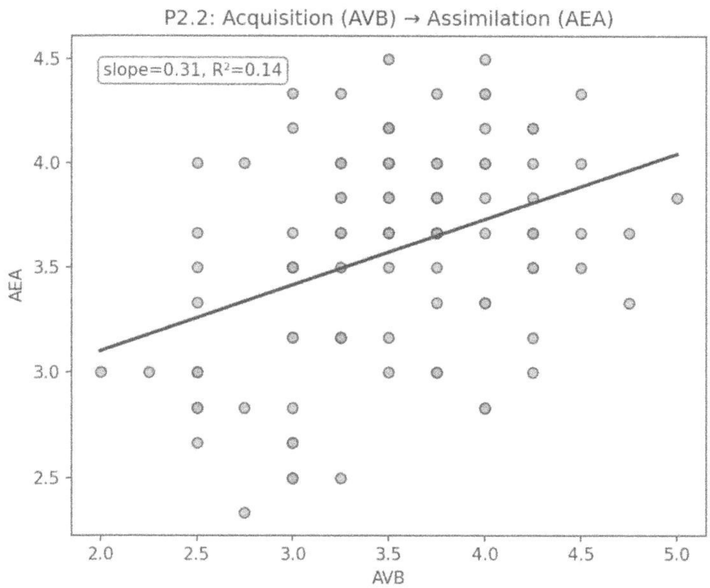

FIGURE 9.5 Impact of acquisition on pedagogical assimilation.

Finally, regarding Proposition P2.2 Acquisition → Assimilation, the data show that the initial acquisition of AI resources for teachers positively affects their assimilation. The model indicates that there is a positive correlation between the variables. The slope is 0.31, while the R^2 is 0.14. This means that approximately 14% of the assimilation is due to the acquisition. This supports the idea that educators who actively seek out and experiment with new AI tools experience tangible benefits, such as reduced effort in providing formative feedback, quicker adaptation of course content, and more efficient redesign of assessments. This is important as it helps to ensure that the daily methods used to teach are in line with the strategic goals that need to be achieved. Essentially, the study illustrates that the acquisition and assimilation of AI tools are key to the transformative process in education.

9.5.5.2 P2.3 – Assimilation → Transformation (Human–AI Workflows, Labs)

Our findings emphasize that assimilation of AI tools and practices by educators is a critical predictor of transformation in the educational environment. This is further emphasized in Model 1, where it is clear that with the assimilation of AI tools, the transformation in the use of human–AI workflows, analytics labs, and in-class supervision is more prominent. This data tells us the story well, showing us that assimilation and transformation go together, but at a rate of 0.52. R^2 value is at 0.28, which means that 28% of the result of the transformation can already be explained by assimilation. However, if we are considering the inclusion of the ECI, this is the measure of the level of support for the initiative. In Model 2, the R^2 value improves to 0.33. This shows that both the process of daily adaptation and the strong support of the institutions push the process forward toward the use of AI-augmented simulations as well as the use of human-in-the-loop designs.

By including the elements in the model, it is clear that good leadership, good communication, and the adoption of sectoral best practices can accelerate the change. The results also confirm Proposition P2.3, again emphasizing that assimilation, when in the right context, is essential in speeding up AI innovation in education.

Figure 9.6 indicates a small yet significant correlation between governance maturity, including transparency, integrity requirements, and equal access, and results such as better decisions and increased learning productivity, indicated by a slope of 0.38 and R^2 of 0.11. This is, of course, not

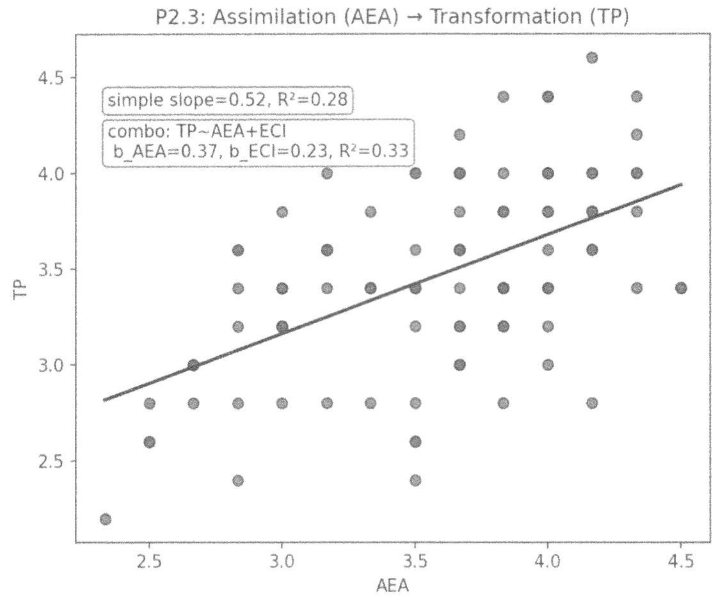

FIGURE 9.6 Governance maturity drives outcome realization.

as high a correlation as assimilation and transformation, yet it indicates that proper governance is indeed effective for improving learning results. The idea is that, although the effect of governance maturity is small, it is effective for safely scaling all efforts and producing actual, quantified results for learning productivity and quality of decisions, which are essential for a smooth rollout of AI systems.

In education, the assimilation transformation predictor is highlighted, especially with the incorporation of human–AI collaboration and analytics labs. This is due to the fact that the accuracy is enhanced with the incorporation of enabling contexts, thus validating Proposition P2.3 that assimilation contexts alone cannot be used to develop simulations for AI with human in the loop.

9.5.5.3 P2.4 – Governance Maturity → Outcome Realization (Safe Scaling)

The research on Proposition P2.4, "*Governance maturity leads to outcome realization*," indicates that high levels of governance practices, such as principles, disclosures, access, and pilot programs, are related to educational outcomes, such as improved learning productivity and better decision quality. Figure 9.7 shows that educators who practice structured GenAI pilot programs (Q32 ≥ 4) have slightly higher outcome scores (3.52 vs. 3.32), supporting the notion that "small-t" risk-slope experimentation is key to safely scaling AI initiatives.

Although these confidence intervals overlap, the trend supports the phased approach to GenAI implementation, as recommended nationally, suggesting that pilot-positive approaches to this implementation do provide benefits, albeit limited in scale. The model also illustrates this trend, with a slope of 0.38 and an R^2 of 0.11, indicating a positive but moderate effect. This illustrates the earlier research propositions discussed, particularly those highlighting the significance of assimilation and transformation in creating change, and that the maturity of governance within these institutions does act as an enabler of positive change and benefits, while also protecting the safe scaling of GenAI implementation, as discussed above.

Figure 9.7, of course, worth noting that this change is likely to result from a number of upstream influences, including the assimilation of AI tools and transformative approaches to teaching, as discussed earlier in this dissertation. Governance maturity is not only protecting safe scaling but is

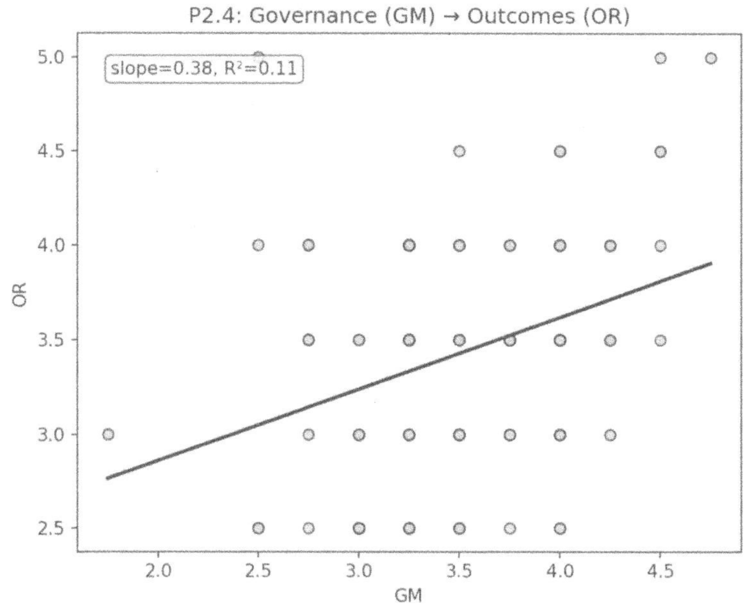

FIGURE 9.7 Governance maturity facilitates outcome realization & safe scaling in AI adoption.

also supporting and acting to enhance the other transformative changes discussed within the above research propositions.

9.5.6 Pilot Practice as a Governance Mechanism

The evidence from adoption patterns provides further support in the examination of the propositions above. Figure 9.8, "Lab Time," shows that nearly all educators enable at least 25% Gen AI lab time, thereby providing support for the proposition above, particularly Proposition P2.3, where assimilated AI practices result in systemic redesign, such as studio-based strategy projects, prompts-to-plans, and oversight. Moreover, positive pilot courses, where $Q32 \geq 4$, have directionally higher outcome scores, where 3.52 is higher than 3.32, with a marginal benefit ($\Delta \approx 0.2$) over no/low pilot courses. Although there is an overlap in the confidence interval, the above is in line with Proposition P2.4, where assimilated small t pilot practices result in safe and effective gains in outcomes. In summary, the above provides support for the overall narrative, where assimilation/ governance maturity is symbiotic in the evolution of AI practices, where pilots result in systemic transformation and effective gains in educational outcomes.

9.5.7 Additional Cross-Construct Perspective

The results from the cross-construct (Figure 9.8) analysis are a strong validation of the ACAP model, where there is a sequential progression from enabling context, to acquisition and assimilation, and then to transformation, governance, and finally outcome realization. As can be seen in Figure 9.9 (Index Means), there is a general decrease in scores from one phase to the next, which is consistent with typical organizational diffusion patterns where foundational enablers are developed before downstream outcomes. These results align directly with the propositions discussed above, where governance and assimilation are key to safe scale and educational practice improvement. Moreover, the barriers and supports identified, such as constraints on instructors' time, training needs, access to licensed AI tools, and templates on governance, mirror exactly the ACAP bottlenecks. This

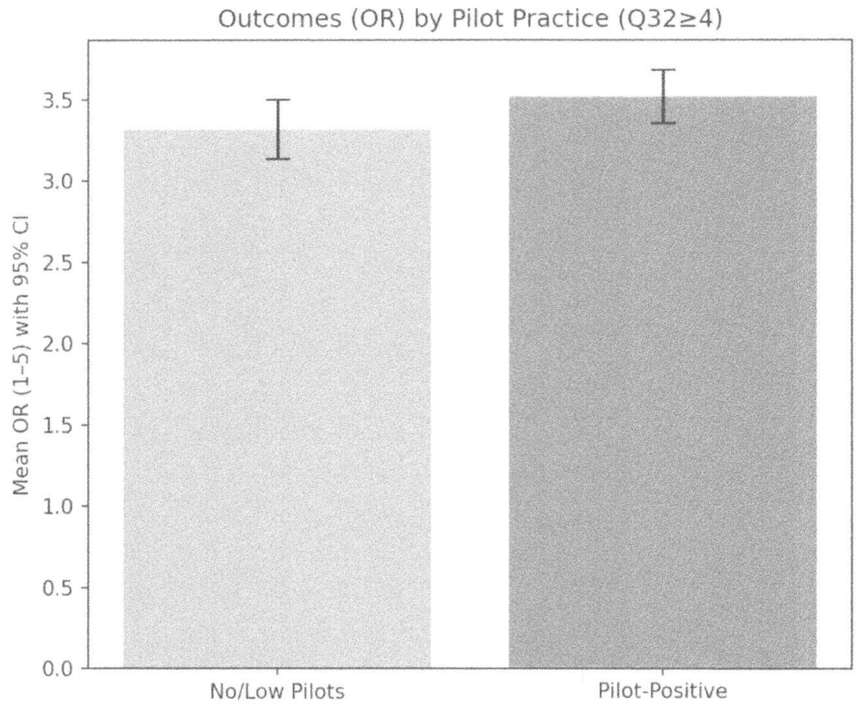

FIGURE 9.8 Outcomes by pilot practice (95% CI).

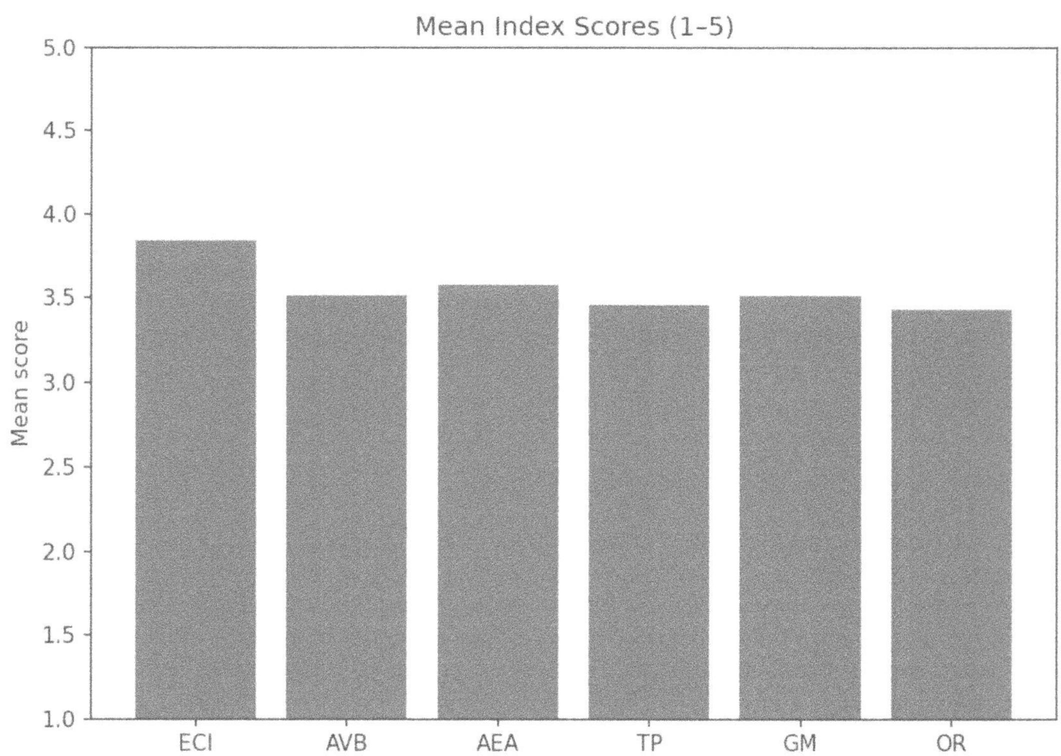

FIGURE 9.9 Mean index scores.

indicates clear areas of opportunity for UK HEIs, especially metropolitan and mixed profile universities, to build on capacity building efforts.

The observed pattern (Figure 9.9), with enabling context leading and realized outcomes trailing, reflects the ACAP maturation pathway noted in the propositions, emphasizing that early enablers are strongest while final outcomes require time for implementation and governance to take effect. This coherent linkage reinforces the broader narrative: that the interplay of assimilation, transformation, and governance maturity is critical for translating AI integration into tangible educational gains, as established in the preceding examination of the research propositions.

9.6 RESEARCH DISCUSSION

Based on the ACAP lens, the UK's integration of GenAI into BS education is best understood as a sequenced capability build: enabling signals foster acquisition; acquisition accelerates assimilation; assimilation enables transformation; and governance converts these gains into exploitation with visible outcomes. In the British sector, national principles (for example, the Russell Group joint statement) and institutional guidance (QAA, Jisc; local university policies) have reduced ambiguity and clarified intent, creating the conditions under which educators can recognize and mobilize the value of external AI knowledge, precisely the dynamic theorized in foundational ACAP work (Cohen & Levinthal, 1990; Jisc, 2024; QAA, 2023, 2024; Russell Group, 2023; Zahra & George, 2002). Building on this, Webster and Westerman's "small t" risk slope perspective adds executional discipline: measure first, build stage capability, and only then scale (Webster & Westerman, 2025).

In terms of *GenAI acquisition*, our data show that it is not mere tool discovery; it is a context-enabled capacity to locate and trial exemplars (cases, prompts, sandboxes) and governance artifacts (disclosure norms, integrity templates) at pace. Where policy scaffolds are salient and consistent, discovery cycles broaden and accelerate, echoing classic ACAP logics of prior knowledge and gatekeeping (Cohen & Levinthal, 1990) and recent UK guidance that explicitly calls for AI literacy, equity of access, and assessment alignment (Bristol BILT, 2024; Jisc, 2024; King's College London, 2025a, 2025b; QAA, 2023, 2024). This helps explain why experimentation, from curated prompt libraries to analytics plug-ins, has diffused rapidly across BS modules in the sample.

Regarding *GenAI Assimilation*, our data show that BS learners transfer knowledge from research to regular teaching practice and assessment. The benefits seen in practice, reduced marginal cost of formative feedback, faster content development, and plausible redesign of assessment with transparency, are in line with recent evidence reviews, which have found GenAI to enhance the timeliness and personalization of feedback while maintaining human review (Xia et al., 2024; Zhao et al., 2024). The UK sector guidance is now focused on authentic assessment, process evidence, and statements about AI use (QAA, 2023, 2024), while university playbooks have implemented these at a program level (Durham University, 2024, 2025; Warwick, 2026). Notably, educators in our sample describe concrete moves, exam-item rehearsal, project pipelines with AI critique loops, and template-based disclosures that turn policy into routine.

The *GenAI transformation* stage takes place after assimilation is established. Instructors orchestrate human–AI workflows and deploy simulations and analytics labs so students can rehearse evidence-based reasoning, sensitivity analysis, and scenario planning in decision environments closer to practice. This hybrid approach aligns with the latest strategy literature on AI as a facilitator of decision (Przegalińska & Jemielniak, 2023) and with UK guidance on operationalizing curriculum-level integration (Bristol BILT, 2024; Jisc, 2024; King's College London, 2025). The risk slope of maturation for AI applications, as described by Westerman, provides insight into the maturation process, with checkpoints for increasing integration and supervision (Webster & Westerman, 2025). The high proportion of lab exposure points to the embedding of the transformation rather than seeing it as an add-on.

In terms of *GenAI exploitation*, our data emphasizes that governance and measurement render capability reliable and repeatable. Institutions codify principles (transparency, parity of access, integrity), normalize pilot-to-scale pathways, and instrument outcomes such as learning productivity (iteration speed, draft quality), decision quality (scenario rigor), and innovation throughput (prototype velocity). This mirrors "realised ACAP," linking transformed practice to performance (Zahra & George, 2002), and is increasingly reflected in UK policy discourse (DfE, 2025), sector guidance (QAA, 2024), and institutional commitments (Queen Mary University of London, 2024, 2025). In our results, pilot-positive educators show directionally higher outcome scores, consistent with small-t logic: "governed experimentation underwrites safe scaling" (Webster & Westerman, 2025).

In terms of *cross-cutting implications*, our data points out six implications. First, the process of measuring transformation can be aided by a "small t" ladder that includes ACAPs for literacy, policy coverage, and training hours (acquisition/assimilation), simulation usage and decision lab throughput (transformation), and reporting of capstone value creation, placements, and KPIs (exploitation) (QAA, 2024; Webster & Westerman, 2025). Second, enhancing the learning process in BS can be achieved by increasing formative feedback and scenario practice, while ensuring integrity and disclosure, now a standard feature in UK playbooks (Bristol BILT, 2024; QAA, 2023, 2024). Enhancing BS learning entails scaling formative feedback and scenario practice while maintaining integrity and disclosure – now standard in UK playbooks (Bristol BILT, 2024; QAA, 2023, 2024). Third, strengthening analytics and decision making requires embedding AI augmented simulations to rehearse forecasting and trade off reasoning, with human in the loop checkpoints (Jisc, 2024; Warwick, 2026). Fourth, transforming applied innovation rests on risk-slope-guided studio work with explicit ethics and governance artifacts (Jisc, 2024; King's College London, 2025a, 2025b). Fifth, the process of reshaping the policy process entails the integration of departmental rules with sector principles such as "permitted/prohibited/required," "disclosure," and "parity of access," which have now become the norm in all the UK HEIs (DfE, 2025; QAA, 2024; Russell Group, 2023). Finally, the voice of educators in the sector has also reflected the journey GenAI has taken to be embedded in the sector's fabric, not as an afterthought. The empirical pattern we observe is coherent with the theory we test *context first*; *acquisition to assimilation*; *assimilation to transformation*; and *governance-enabled exploitation* that sustains impact at scale. In short, this is ACAP made operational for BS education in the UK, with clear lines of sight from sector principles and small t pilots to robust, ethical, and performance-relevant practice

9.7 CONCLUSION

In this chapter, the complex role of GenAI in the development of absorptive capacity in the BS education domain in the universities of Britain has been explored through the use of both theoretical frameworks and empirical research. From the theoretical perspective, the present study extends the ACAP model from the field of organizational learning (Abou-Foul et al., 2023; Cohen & Levinthal, 1990) to the field of higher education, thus underlining the importance of GenAI in the process of knowledge acquisition, assimilation, transformation, and exploitation. Through the mapping of the stages of the absorptive capacity model, the current study provides an insightful framework for the role of institutional signals, policies, and practices in empowering educators in the domain of AI, thereby ensuring that the use of AI is not just an upgrade but an actual transformation in the domain of education.

From an empirical perspective, the chapter draws upon a wide-ranging survey carried out among 100 BS educators from a range of different British universities. The results show that having institutional support can play a significant role in helping educators make use of GenAI technology, which can improve the assimilation rates. The results also show that acquisition can play a role in pedagogical assimilation, which can improve the efficiency in redesigning. Perhaps most interestingly, the results support the ACAP sequence in full, from enabling context to acquisition to assimilation, which in turn drives transformational human–AI workflows and, finally, outcomes grounded in developed governance structures. Yet again, there are challenges to be found in these results, including those relating to resource constraints and redesigning assessments.

With a view to the future, the following recommendations are made. First, universities need to invest in a comprehensive training infrastructure so that educators, not just early adopters, can meaningfully interact with GenAI. Second, there is a need to prioritize and enhance the development of comprehensive governance structures. Third, the conduct of pilot projects should be encouraged, creating a "safe space" where one can experiment, where learning can take place. Fourthly, future research should be conducted using longitudinal research designs that will seek to address the comprehensive impacts of GenAI on teaching and learning, including issues of reliability, which were evident in the measurement approach used for the study. In essence, this chapter is a plea to humanize the development of GenAI in BS education. Contrary to the view of AI as an educator replacement, the findings of the study have shown that, when well-integrated, AI is likely to greatly improve the creative and strategic capabilities of the educator, as well as the learner. It is the relationship between man and AI, imbued with the elements of ethics, critical thinking, and learning, that is likely to shape the future of business education.

REFERENCES

Abou-Foul, M., Ruiz-Alba, J. L., & López-Tenorio, P. J. (2023). The impact of artificial intelligence capabilities on servitization: The moderating role of absorptive capacity-A dynamic capabilities perspective. *Journal of Business Research, 157*, 113609. Available at: https://doi.org/10.1016/j.jbusres.2022.113609

Bristol Institute for Learning and Teaching. (2024). *Guidance on AI*. University of Bristol. Available at: https://www.bristol.ac.uk/bilt/sharing-practice/guides/guidance-on-ai/

Butarbutar, R. (2025). Review article on Best practices: Leveraging AI tools in teaching business and economics education. *Pedagogical Constellation, 4*(2), 256–277. Available at: https://doi.org/10.69821/constellations.v4i2.118

Cambridge Judge Business School. (2025). *Step into AI-driven business and leadership learning*. Available at: https://www.jbs.cam.ac.uk/2025/step-into-ai-driven-business-and-leadership-learning/

Cambridge Judge Business School. (2026). *Cambridge AI Leadership Programme* (programme page). Available at: https://www.jbs.cam.ac.uk/executive-education/leadership/cambridge-ai-leadership-programme/

Chan, C. K. Y. (2023). Generative artificial intelligence (AI) in higher education. *Computers and Education: Artificial Intelligence, 5*, 100197. Available at: https://doi.org/10.1016/j.caeai.2023.100197

Cohen, W. M., & Levinthal, D. A. (1990). Absorptive capacity: A new perspective on learning and innovation. *Administrative Science Quarterly, 35*(1), 128–152. Available at: https://doi.org/10.4324/9780080517889

Department for Education. (2025, August 12). *Generative artificial intelligence (AI) in education*. GOV.UK. Available at: https://www.gov.uk/government/publications/generative-artificial-intelligence-in-education/generative-artificial-intelligence-ai-in-education

Durham University (DCAD). (2024, July 16). *Generative AI in learning, teaching & assessment*. https://dcad-resourcebank.webspace.durham.ac.uk/2024/07/16/generative-ai-in-learning-teaching-assessment/

Durham University (DCAD). (2025, August 27). *Assessment and marking in light of generative AI*. https://dcad-resourcebank.webspace.durham.ac.uk/2025/08/27/assessment-and-marking-in-light-of-generative-ai/

Dutta, S., Ema, U. P., Sekhar, R. P., & Arivazhagan, R. (2025a). AI-driven transformation in education business: A conceptual exploration of solutions to complex industry issues. *International Business & Economics Studies, 7*(4), 101. Available at: https://doi.org/10.22158/ibes.v7n4p101

Dvouletý, O., Jirásek, M., Giglio, C., & Liguori, E. (2025). Artificial intelligence, digitalization, and new technologies in shaping business and management research. *Cogent Business & Management, 12*(1), 2580126. Available at: https://doi.org/10.1080/23311975.2025.2580126

Jisc. (2024, June 11). *New toolkit for colleges and universities: AI maturity model*. Available at: https://www.jisc.ac.uk/news/all/new-toolkit-for-colleges-and-universities-sets-pathway-for-effective-adoption-of-ai

King's College London. (2025a). *Macro-level guidance: University-wide principles and policy (aligning with RG)*. Available at: https://www.kcl.ac.uk/about/strategy/learning-and-teaching/ai-guidance/macro-level

King's College London. (2025b, April). *King's guidance on generative AI for teaching, assessment and feedback*. Available at: https://www.kcl.ac.uk/about/strategy/learning-and-teaching/ai-guidance

Mohamad, M., Balıkel, A. E., Chowdhury, D., Russell, C., & Mahmoud, S. (2026): Artificial intelligence for business model innovation in digital sustainability companies: Multiple cases from the Netherlands. *World Journal of Entrepreneurship, Management and Sustainable Development (WJEMSD), 22*(1–2), 109–125. Available at: https://doi.org/10.47556/J.WJEMSD.22.1-2.2026.7

Mohamad, M., Chowdhury, D., Russell, C., Balikel, A., & Kawalek, P. (2025). How GenAI transforms computer engineering education: The case of the middle East and North Africa. *Journal of Engineering Education Transformations, 39*(3), 3–19. Available at: https://doi.org/10.16920/jeet/2025/v39is1/25129

Pang, W., & Wei, Z. (2025). Shaping the future of higher education: A technology usage study on generative AI innovations. *Information, 16*(2), 95. https://doi.org/10.3390/info16020095

Prado, P. F., & Mantovani, D. (2025). Generative AI: Capturing strategic value in businesses. *Revista de Gestao, 32*(3), 208–223. Available at: https://doi.org/10.1108/REGE-02-2025-0026

Przegalińska, A., & Jemielniak, D. (2023). *Strategizing AI in business and education: Emerging technologies and business strategy*. Cambridge University Press. https://doi.org/10.1017/9781009243520

Quality Assurance Agency for Higher Education (QAA). (2023, May 8). *Maintaining quality and standards in the ChatGPT era*. https://www.qaa.ac.uk/docs/qaa/members/maintaining-quality-and-standards-in-the-chatgpt-era.pdf

Quality Assurance Agency for Higher Education (QAA). (2024, February). *Quality Compass: Navigating the complexities of the AI era in HE*. https://www.qaa.ac.uk/docs/qaa/news/quality-compass-navigating-the-complexities-of-the-artificial-intelligence-era-in-higher-education.pdf

Queen Mary University of London. (2024). *University AI principles, guidance and policies*. Available at: https://www.qmul.ac.uk/queenmaryacademy/educators/innovation-and-scholarship/innovative-pedagogies/centre-for-excellence-in-ai-in-education/resources/university-guidance-and-policies/

Queen Mary University of London. (2025). *University AI principles, guidance and policies*. Available at: https://www.qmul.ac.uk/queenmaryacademy/educators/innovation-and-scholarship/innovative-pedagogies/centre-for-excellence-in-ai-in-education/resources/university-guidance-and-policies/

Russell Group. (2023, July 4). *Russell Group principles on the use of generative AI tools in education*. Available at: https://www.research.pitt.edu/sites/default/files/russell_group.pdf

Singh, V., Hughes, L., Albashrawi, M. A., Jeon, I., & Dwivedi, Y. K. (2026). Generative artificial intelligence (GenAI) in procurement and supply chain management: Applications, opportunities and challenges. *Journal of Systems and Information Technology, 28*(1), 123–144. https://doi.org/10.1108/JSIT-03-2025-0139

Strielkowski, W., Grebennikova, V., Lisovskiy, A., Rakhimova, G., & Vasileva, T. (2025). AI-driven adaptive learning for sustainable educational transformation. *Sustainable Development, 33*(2), 1921–1947. Available at: https://doi/10.1002/sd.3221

University of Bristol—BILT. (2024). *Our commitment to the Russell Group principles*. Available at: https://www.bristol.ac.uk/bilt/sharing-practice/guides/guidance-on-ai/our-commitment-to-the-russell-group-principles/

Webster, M., & Westerman, G. (2025). Generate value from GenAI with "Small t" transformations. *MIT Sloan Management Review (Online)*, 15-21. Available at: https://www.proquest.com/openview/108b6a9f2480fcc14e538ba7449cf9ba/1?pq-origsite=gscholar&cbl=6831990

Xia, Q., Weng, X., Ouyang, F., Lin, T. J., & Chiu, T. K. F. (2024). A scoping review on how generative AI transforms assessment in higher education. *International Journal of Educational Technology in Higher Education, 21*(40). Available at: https://doi.org/10.1186/s41239-024-00468-z

Zahra, S. A., & George, G. (2002). Absorptive capacity: A review, reconceptualization, and extension. *Academy of Management Review, 27*(2), 185–203. Available at: https://www.jstor.org/stable/4134351

Zhao, J., Chapman, E., & Sabet, P. G. P. (2024). Generative AI and educational assessments: A systematic review. *Educational Research and Practice Journal, 51*, 124–155. Available at: https://search.informit.org/doi/10.3316/informit.T2025061800005500592641426

10 Act for an AI School

Smart Diagnostics and Personal Support

Nadia Elassassi, Said Machwate, Rachid Bendaoud, and Said El Hasri

10.1 INTRODUCTION

Teaching at the right level (TaRL) is an educational method ameliorated by the organization Pratham, whose objective is to adapt teaching to the real level of the learners rather than their age or class (Jazuli, 2022). Africa and Asia, documented in many countries in Africa and Asia, is based on regular diagnostic evaluation, the temporary grouping of students according to their skills, and the use of targeted remediation activities. The report Pilot of a TaRL-based approach in Lebanon (2025) demonstrates that applying this approach significantly improves students' achievements and contributes to reducing the persistent learning gaps accounted for heterogeneous classes (World Bank, 2025).

According to UNICEF, TaRL is a particularly suitable intervention for low-resource settings, as it requires neither sophisticated equipment not costly infrastructure (UNICEF, 2024).

Using simple and frequent assessments, TaRL allows students to be grouped according to their real achievements, which develops targeted and adapted teaching. This strategy has confirmed its effectiveness in development foundational skills in several low- and middle-income countries (Banerjee et al., 2016). Today, the advances of Artificial Intelligence (AI) constitute an essential lever to strengthen and empower this teaching philosophy. For its ability to analyze large amounts of data and propose personalized recommendations, AI makes it easier to determine the true level of each learner and build adapted learning paths (UNESCO, 2024).

The ways to integrate AI into the education system are diverse and aim to address current challenges, including diversity in learning levels and the excessive load on teachers (Coppin, 2025; Joshi, 2023).

First, personalized learning, based on adaptive algorithms, is one of the most popular applications of AI. Platforms such as Knewton or DreamBox define the timely analysis of learner data for learning content with their specific capabilities (Siddiqui et al., 2025). The results prove that this approach reduces disparities between learners, allowing everyone to develop at their own rhythm.

Second, Intelligent Tutoring Systems (ITS) provide interactive assistance to users in STEM subjects (science, technology, engineering, and mathematics). These virtual teachers identify difficulties based on learner performance and provide direct explanations for improving their level. For educational environments where access to teachers is limited, these systems have shown significant improvements in learning outcomes (Luckin et al., 2016).

Third, automating administrative tasks is another major application of AI. AI saves time by correcting homework, organizing school calendars, or even submitting detailed reports. This time can enable teachers to devote themselves more to pedagogical support and human interactions, which remain at the heart of the educational act (Chassignol et al., 2018).

The studies have confirmed that the use of AI not only improves learners' outcomes but also improves their motivation and dedication to the subjects studied. Therefore, the impacts on learning

DOI: 10.1201/9781003767855-10

are multiple and measurable. As a matter of fact, the study by Holmes et al. (2019) explains that AI allows teachers to better track each learner's progress, and more easily identify specific difficulties that need to be faced. Moreover, interactive and personalized learning increases learner engagement, especially those who find it difficult to follow traditional courses (Holmes et al., 2019).

However, it is essential to ensure the responsible use of learner data, ensure confidentiality, and avoid algorithmic biases that can create inequality. Integrating AI into education raises ethical and technical concerns. Zawacki-Richter et al. (2019) emphasize the need to train teachers in AI technologies so that they can use them in their educational practices efficiently and ethically (Zawacki-Richter et al., 2019).

10.2 THEORETICAL CONTEXT

Depending on the group of learners, classes are classified as homogeneous or heterogeneous. Homogeneous classes bring together learners with similar levels, making it easier for teachers to teach them more effectively. As for the heterogeneous classroom grouping between learners from different levels, which encourages cooperative learning and community support (Dupriez & Draelants, 2004).

Diagnostic assessment is an essential step in understanding learners' level of knowledge at the beginning of a learning period. Its goal is to identify prior knowledge, detect gaps and adapt teaching to the specific needs of learners (Black & Wiliam, 1998). Very known used methods include written tests, questionnaires, and classroom observations, permitting teachers to plan targeted pedagogical interventions. After a diagnostic test, it is essential to support students to help them succeed. Teachers can execute targeted interventions, forming support groups for learners with similar needs. Regular constructive comments help students grasp their errors and progress effectively (Dupriez & Draelants, 2004).

Heterogeneous classes often present specific challenges concerning the diversity of learners' skill levels. In these classes, it is well known to find great variation in students' knowledge and skills, making it difficult to adapt to a single pedagogical approach that is effective for all. Some students may have real gaps, while others have already learners the concepts taught (GHERBAOUI, 2017).

This unbalance makes the task of teachers harder, they must find strategies to meet the varied needs of individual learners while maintaining a pace of progress for the whole class (Dupriez & Draelants, 2004). To overcome these difficulties, supporting learners in a heterogeneous class requires differentiated and individualized pedagogical interventions. Teachers can put support groups for learners with similar skills, to offer them a more targeted and adapted environment to their needs.

Finally, constructive feedback is an essential element of support in heterogeneous classes. By providing frequent feedback, teachers help students understand their mistakes and improve strategies for overcoming them. This approach encourages learner engagement and motivation by demonstrating that their efforts are recognized and appreciated, which is particularly important in a context where ability levels vary (Black & Wiliam, 1998).

AI can be used to personalize learning paths when it becomes difficult for teachers to meet the needs of all students, because AI offers innovative solutions to the challenges faced by heterogeneous groups after diagnostic assessment. For example, AI-based systems can analyze the results of diagnostic assessments and suggest exercises adapted to each learner's proficiency level, such as promoting differentiated learning (Gupta, 2024; Halkiopoulos & Gkintoni, 2024; Hariyanto et al., 2025; Mario et al., 2025).

AI proposes activities that help learners progress at their own pace by targeting specific gaps (Woolf, 2008). AI systems can analyze learners' responses in real time, identify errors, and provide detailed explanations as well as appropriate corrective actions. One of its most important uses is its ability to provide immediate and continuous feedback to learners, a task that poses a challenge for teachers in a heterogeneous classroom environment. This ability to quickly adapt teaching to individual needs improves the effectiveness of pedagogical support and keeps learners engaged in their learning (Luckin et al., 2016). Furthermore, AI as a predictive system can identify potential barriers in advance and modify teaching strategies to improve learners' chances of success, especially in classrooms with different skill levels. This is done by creating easier individual educational paths by using data to customize educational content to suit each learner's skills and learning styles. By making teaching more responsive and individualized, AI helps promote more equal and inclusive learning for all students (UNESCO, 2019).

10.3 METHODOLOGY

In this study, we relied on the Correct Level Teaching (TaRL) curriculum, which has already been tried in many leading schools in the Moroccan education system. It is based on determining the true level of each learner, regardless of the class or formal program to which he belongs, and adapting teaching to the skills he masters. It therefore aims to go beyond the traditional logic based solely on curriculum progress, to prioritize learning based on students' real needs.

To implement this study incorporated the use of an AI platform learning (AI). The objective was to rely on a tool capable of generating accurate, rapid, and reliable diagnosis, while allowing individual monitoring of each learner's progress. The use of AI-based technology is part of the perspective of modernizing teaching practices, by offering functions that are impossible to fill within a strict manual framework, especially with regard to data processing, customization, and adaptability.

To achieve this goal, each student from the three departments in the first year of middle school took an initial diagnostic test via the Kwyk.fr platform. This tool was chosen based on several methodological and practical considerations. On the one hand, Kwyk is one of the few accessible platforms that offers a free one-month training period, allowing for full experimentation without financial restrictions. On the other hand, the platform allows the registration of an unlimited number of students, which facilitates the formation of a representative sample according to study requirements.

In addition, Kwyk offers a range of features that are particularly suitable for an educational differentiation approach. The platform's characteristics focus on making it a relevant methodological tool for conducting reliable initial diagnostics, organizing differentiated teaching, and ensuring continuous monitoring based on real and up-to-date data. This is done by automatically sending the detailed results of each student to the registered teacher, which facilitates the processes of data collection, analysis, and archiving

As the platform is equipped with an infinite number of exercises, it automatically generates activities that adapt to the actual level of each learner. This allows the student to retake the test several times, providing various exercises on each attempt but equivalent in difficulty. This mechanism promotes intensive and gradual processing until complete mastery of the evaluated concept. This diagnostic and adaptive phase was conducted in October 2025 with a sample of learners distributed across three classes at the same public located in Marrakesh, Morocco, ensuring precise allocation of learning paths.

Students have free access to the platform during the trial month, while simultaneously quantitatively tracking sessions (number of tests, pass rate, and time spent), which are then collected for further analysis of individual and group learning gains. Once the diagnostic test is completed, the AI platform ensures a repetitive and independent learning loop: with each failure or resumption, new exercises of the same level are suggested, with immediate feedback to enhance understanding. A student only progresses when mastery is validated by repeated successes, which then leads to an automatic transition to the next idea. This dynamic adaptation, made possible by AI, avoids the frustration of extremely difficult exercises and the boredom of identical repetitions, while enhancing independence through self-assessment and visual progress indicators.

A diagnostic post-test measures progress at the end of the month, compared to initial results, to evaluate the effectiveness of this hybrid TaRL-AI methodology, with indicators such as the evolution of average scores.

10.4 RESULT AND DISCUSSION

Statistical analysis of data from a sample of 55.24% boys and 44.76% girls aged 11 to 13 years. Grades obtained range from a minimum of 2 to a maximum of 19, with an overall average of 10.1. There is a slight difference between the 2 groups: boys get 9.7 points, while girls get 11.0 points. This trend indicates a slight advantage for female students in mastering the mathematical skills assessed. The distribution of scores distinguishes between three categories of performance: 28 learners (22.6%) have a low level, and their scores range from 2 to 6; 42 learners (33.9%) are at the intermediate level, and their scores range from 7 to 12; 30 learners (24.2%) reach a high level, and their scores range from 13 to 19.

The distribution shown in Figure 10.1 highlights the heterogeneity of levels within the categories studied. It more precisely targets students who belong to the low-performing group. This allows students to identify learner profiles that require interventions.

Through this diagnostic rule, learners with low scores were exposed to more substantial exercises, while learners with intermediate or high scores worked on more complex activities. Data analysis shows that students engaged in independent work, characterized by strong motivation and several successive attempts at each exercise. The process of iteration and self-correction supported by immediate feedback has contributed to the gradual improvement in its performance. Notably, students in the categories "low" and "medium" showed the most significant progress, confirming the relevance of adapting the level of requirement to initial skills.

Table 10.1 summarizes the performance profile of students by class, showing the overall level of proficiency, dispersion of performances, commitment, and strong or weak skills. Class A is homogeneous and efficient, requiring targeted remediation. Class B presents moderate dispersion with varied profiles, justifying a gradual consolidation. Class C is the most heterogeneous, with significant gaps and contrasting commitment, requiring pedagogical differentiation. This table thus guides the intervention priorities for each class.

The statistics presented in Figure 10.2 clearly identify students' difficulties according to the different concepts addressed. For several themes, such as numerical sequences or arithmetic and geometric sequences, the results show that students do not always succeed on the first try, but progress significantly on the second attempt, which reflects the effectiveness of the remediation work.

This visual diagnosis highlights areas that require enhanced support, while demonstrating concepts that students are able to quickly establish. Secret modeling concepts, particularly exponential growth, show a very high success rate on the second attempt and few errors, indicating good customization after debugging. This level of progress reflects not only a significant commitment but

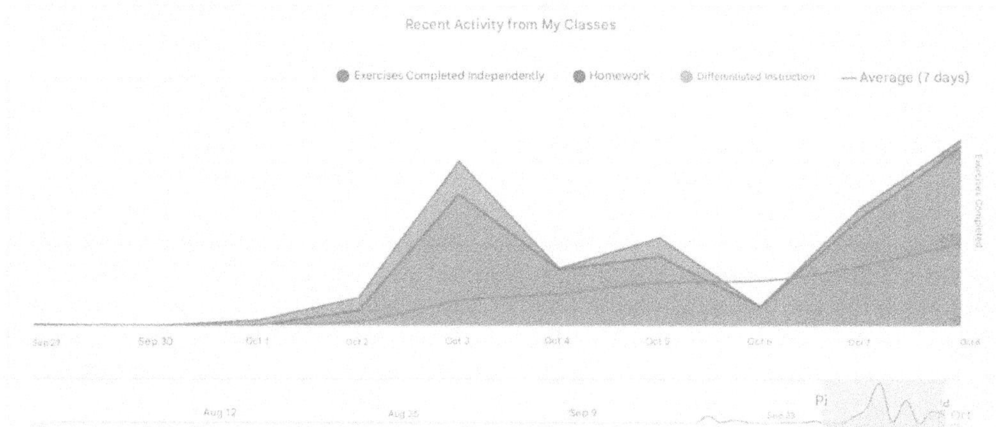

FIGURE 10.1 Progression of the students in autonomous exercises and on duty.

TABLE 10.1
Diagnostic Performance Profile and Educational Priorities by Class

Class	Overall Level of Proficiency	Dispersion of Performance	Student Engagement	Strong Skills	Fragile Skills	Educational Priority	Details and Justification
A	High ($M = 16.0$)	Low to moderate	High (targeted attempts)	Exercises 3621, 5840, 999	Exercises 631, 633, 720	Targeted conceptual remediation	Majority of grades ≥14. The spread is limited: some occasional failures, but most students succeed well. The attempts are numerous but focused on the same students or types of exercises.
B	Medium-high ($M = 15.5$)	Moderate	Moderate to high	Exercises 5840, 3621, 292	Exercises 633, 631, 720	Gradual consolidation	Greater diversity of intermediate grades. Dispersion is installed: the distribution of scores is more varied, indicating heterogeneous learning profiles and less regularity in attempts.
C	Medium-high ($M = 15.8$)	Moderate to high	High (repeated attempts)	Exercises 3621, 5840, 999	Exercises 633, 292, 720	Differentiated instruction	The dispersion is marked and structuring: gaps between students are significant, and attempts show very contrasting engagement behaviors.

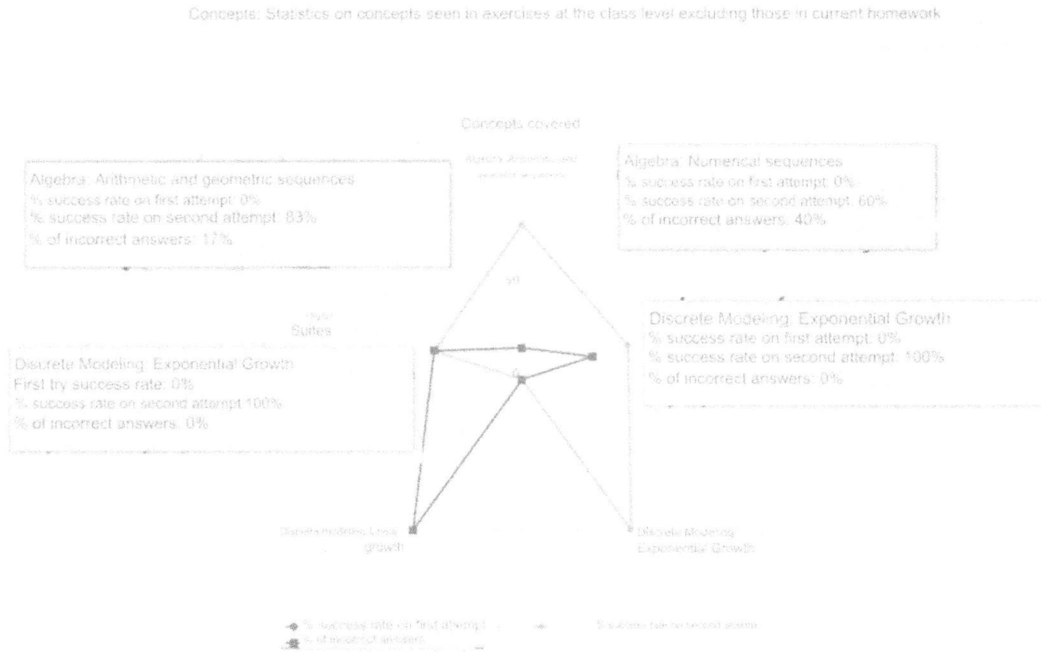

FIGURE 10.2 Trial and error statistics by category of concepts.

also a genuine willingness on the part of learners to invest in their education. Practical activities and exercises play a crucial role in guiding learners according to the work of Venet et al. (2017) in the field of proximal development, through tasks that are slightly beyond their current abilities, supported by appropriate resources (Venet et al., 2017). This allows learners to have a deeper understanding of concepts while gradually improving their skills.

To promote active learning and skills development, this high level of engagement is essential. According to self-determination theory, research that providing opportunities, for learners to succeed in exercises, enhances their intrinsic motivation, and encourages them to persevere and overcome the challenges they face in their learning. This motivation is a key driver of academic success, especially in mathematics, where continuous practice is required to master concepts (Paumier & Chanal, 2018).

As a subject of study, mathematics requires regular practice so that learners can truly master the concepts, the next way to learn mathematics is to solve problems, which requires the active and regular participation of learners. The exercises allow learners to discover techniques, improve their strategies, and develop critical thinking. The more students train, the more confident they become in their mathematical skills, thus enhancing their ability to handle complex problems with confidence (Coppé & Dorier, 2024).

Immediate feedback and repetition of exercises are critical to improving learners' performance in mathematics, according to Kuncel et al. (2005). These methods provide opportunities to correct and understand learners' mistakes, and develop an analytical approach to problem solving. To maintain learner engagement and promote continuous progress, this process of reinforcement and modification is essential.

AI has proven effective in engaging learners' and improving performance. AI not only improves learner outcomes, as highlighted in the study by Zawacki-Richter et al. (2019), but also keeps them motivated through interactive and engaging environments (Zawacki-Richter et al., 2019). The platform, powered by AI algorithms, has helped generate learners' interest to make learning activities more engaging, according to the findings of Popenici & Kerr (2017), which analyzes the impact of AI on student engagement in education (Popenici & Kerr, 2017).

To design exercises according to the needs of learners, it is recommended to use AI because it allows for accurately targeting their gaps and increasing overall performance, as confirmed by Mayer-Schönberger and Cukier (2013). Holmes et al. (2019) also support this personalization of learning, highlighting the transformative role of AI in education, facilitating accurate tracking and continuous adaptation of educational practices (Holmes et al., 2019).

10.5 CONCLUSION

The results of this study show strong evidence of the efficiency of integrating AI into personalized learning tools, by improving the accuracy of responses, as well as students' conceptual understanding and cognitive engagement, based on a detailed analysis of individual needs. This is what was stated in research conducted by Luckin et al. (2016), which proved the contribution of AI technologies to transforming educational practices and improving learning.

The instant feedback provided by the platform is an essential mechanism to support self-regulation and promote self-correction. Adaptive and gradual repetition of exercises also contributes to enhancing achievements gradually and permanently, while interaction and diversity of activities stimulate motivation, which is a crucial factor in success in mathematics. Thus, the educational levers provided by AI have allowed for consistent and measurable progress in line with educational expectations.

Therefore, AI tools are considered a means of support and not a substitute for the human educational role. Therefore, it is essential to develop continuous training programs that enable teachers to effectively master the use of AI-based tools, understand their algorithmic logic, and critically and appropriately integrate them into their classroom practices. This approach would not only ensure the informed and ethical use of AI, but also enhance the teacher's central role in mediating, supporting and interpreting learning.

The study poses several methodological limitations that should be emphasized despite the results obtained. First, the relatively small sample size limits the range of possible generalizations to other educational contexts or student groups. Second, the time period of the study does not allow assessing the long-term impact of AI on enhancing skills nor examining the sustainability of learning after the experimentation phase.

In addition, we were unable to fully control some contextual factors: such as the physical conditions for having access to the platform, the educational environment, the specific practices of each learner or even classroom dynamics. These variables, which are likely to influence observed performance, constitute important limits to interpreting the results.

To reinforce these results, future studies should be based on broader, more diverse and representative samples, as well as longer periods of experiments, making it possible to evaluate the development of learning in the medium and long term. It would also be appropriate to adopt a systematic approach that integrates AI into a global educational ecosystem that combines teaching practices, school organization, digital resources and teacher training.

Finally, exploring the impact of AI on cross-cutting skills such as critical thinking, problem solving, mathematical creativity, or even attention is a promising way to better understand the role of these technologies in sustainable and equitable educational transformation that focuses on the real needs of learners.

REFERENCES

Banerjee, A., Banerji, R., Berry, J., Duflo, E., Kannan, H., Mukherji, S., Shotland, M., & Walton, M. (2016). *Mainstreaming an effective intervention evidence from randomized evaluations of "teaching at the right level" in India.* https://doi.org/10.3386/w22746

Black, P., & Wiliam, D. (1998). Assessment and classroom learning. *Assessment in Education: Principles, Policy & Practice*, *5*(1), 7–74. https://doi.org/10.1080/0969595980050102

Chassignol, M., Khoroshavin, A., Klimova, A., & Bilyatdinova, A. (2018). Artificial Intelligence trends in education: A narrative overview. *Procedia Computer Science*, *136*, 16–24. https://doi.org/10.1016/j.procs.2018.08.233

Coppé, S., & Dorier, J.-L. (Eds.). (2024). *La résolution de problèmes en mathématiques: Enjeux pour l'enseignement et l'apprentissage*. UGA Éditions, EDP Sciences. https://doi.org/10.4000/12ct8

Coppin, B. (2025). The potential of AI in education: Personalizing learning. *International Journal of Artificial Intelligence for Science (IJAI4S)*, *1*(2). https://doi.org/10.63619/ijai4s.v1i2.003

Dupriez, V., & Draelants, H. (2004). Classes homogènes versus classes hétérogènes: Les apports de la recherche à l'analyse de la problématique. *Revue française de pédagogie*, *148*(1), 145–165. https://doi.org/10.3406/rfp.2004.3258

Gherbaoui, A. (2017). *L'hétérogénéité des apprenants en classe de F L E : Un défi à relever.*

Gupta, T. (2024). Adaptive learning systems: Harnessing AI to personalize educational outcomes. *International Journal for Research in Applied Science and Engineering Technology*, *12*(11), 458–464. https://doi.org/10.22214/ijraset.2024.65088

Halkiopoulos, C., & Gkintoni, E. (2024). Leveraging AI in e-learning: Personalized learning and adaptive assessment through cognitive neuropsychology—a systematic analysis. *Electronics*, *13*(18), 3762. https://doi.org/10.3390/electronics13183762

Hariyanto, Kristianingsih, F. X. D., & Maharani, R. (2025). Artificial intelligence in adaptive education: A systematic review of techniques for personalized learning. *Discover Education*, *4*(1), 458. https://doi.org/10.1007/s44217-025-00908-6

Jazuli, L. (2022). Teaching at the right level (TARL) through the all smart children approach (sac) improves student's literature ability. *Progres Pendidikan*, *3*(3), 156–165. https://doi.org/10.29303/prospek.v3i3.269

Joshi, M. (2023). Adaptive learning through artificial intelligence. *SSRN Electronic Journal*. https://doi.org/10.2139/ssrn.4514887

Kuncel, N. R., Credé, M., & Thomas, L. L. (2005). The validity of self-reported grade point averages, class ranks, and test scores: A meta-analysis and review of the literature. *Review of Educational Research*, *75*(1), 63–82. https://doi.org/10.3102/00346543075001063

Luckin, R., Holmes, W., Griffiths, M., & Corcier, L. B. (2016). *Intelligence unleashed: An argument for AI in education*. Pearson.

Mayer-Schönberger, V., & Cukier, K. (2013). *Big data: A revolution that will transform how we live, work, and think*. Houghton Mifflin Harcourt.

Paumier, D., & Chanal, J. (2018). Motivation autodéterminée, conséquences motivationnelles et hypothèse de spécificité en contexte scolaire: État des lieux, implications et perspectives. *Revue française de pédagogie*, *203*, 111–130. https://doi.org/10.4000/rfp.8160

Popenici, S. A. D., & Kerr, S. (2017). Exploring the impact of artificial intelligence on teaching and learning in higher education. *Research and Practice in Technology Enhanced Learning*, *12*(1), 22. https://doi.org/10.1186/s41039-017-0062-8

Siddiqui, M. T., Mansoori, M. V., Siddiqui, M. A., & Yadav, A. (2025). AI-enabled pedagogy: Advancing education through innovative teaching tools and the AI-TEACH model. *Journal of Informatics Education and Research*, *5*(1). https://doi.org/10.52783/jier.v5i1.2261

UNESCO. (2019). *Artificial Intelligence in Education: Challenges and Opportunities for Sustainable Development*. https://unesdoc.unesco.org/ark:/48223/pf0000366994

UNESCO. (2024). *Rapport mondial de suivi sur l'éducation, 2023: Les technologies dans l'éducation : qui est aux commandes?* GEM Report UNESCO. https://doi.org/10.54676/ZGKD9418

UNICEF. (2024). *Lessons Learned from Teaching at the Right Level (TaRL) in MIA Interventions*. https://www.unicef.org/lac/en/reports/lessons-learned-teaching-right-level?

Venet, M., Correa Molina, E., Nootens, P., & Roberge, M. (2017). La zone proximale de développement: Une zone de changements intérieurs pour les étudiantes du baccalauréat en adaptation scolaire et sociale?1. *Nouveaux cahiers de la recherche en éducation*, *19*(1), 88–110. https://doi.org/10.7202/1040664ar

Woolf, B. (2008, September 9). *Building Intelligent Interactive Tutors: Student-centered Strategies for Revolutionizing E-learning*. https://www.semanticscholar.org/paper/Building-Intelligent-Interactive-Tutors%3A-Strategies-Woolf/3056591702e3b34baf579d7bf8175408d274be46

World Bank. (2025). *Pilot of a Teaching at the Right Level (TaRL)-Based Approach in Lebanon*. World Bank.

Zawacki-Richter, O., Marín, V. I., Bond, M., & Gouverneur, F. (2019). Systematic review of research on artificial intelligence applications in higher education – Where are the educators? *International Journal of Educational Technology in Higher Education*, *16*(1), 39. https://doi.org/10.1186/s41239-019-0171-0

11 AI-Enabled Education

Implications for Teaching, Learning, and Assessment

Kamini Bhasin

11.1 INTRODUCTION

The impact of artificial intelligence (AI) on learning today is profound. It is one of several key advancements in technology that is changing the way we learn and teach in our education system. Previously, we have supported our teachers/students by using tools such as an online learning platform, videos, and interactive software. However, now AI is changing all of that. With AI, computers can now process, analyze, generate, and provide results and make decisions with little or no human input. The U.S. Department of Education (2023) states, "The role of AI in education has progressed from simple data collection to more complex tasks such as providing personalized feedback, providing recommendations for personalized learning paths, and tracking student performance over time." Through these developments, there is little doubt that AI can change the way we teach or how our students receive instruction.

The Department of Education (2023) emphasizes that without thoughtful guidance AI may unintentionally widen equity gaps or undermine essential human judgment in educational decision-making. As a large number of academicians now use AI tools in routine professional tasks. The most prominent examples include use of generative text tools for creating lesson plans, automated feedback systems, language-support applications for multilingual learners, and adaptive learning platforms that monitor student progress.

Using AI tools may save time. It also offers flexibility in diverse learning needs. Academicians express legitimate concerns regarding the use of AI and the accuracy of AI-generated outputs. They show their apprehension toward the possibility of reinforcing biases embedded in training data, risks to student privacy, and the potential misuse of surveillance-oriented technologies in classrooms.

AI can help educators in many ways. It can increase access to high-quality teaching resources. It can support teachers' workloads and increase opportunities for personalized learning. But we have to ensure that it is developed ethically and contextualized for actual classroom settings. The development of AI has created an urgency to create strategies for educational institutions to ensure tool usage. It has increased the demand of accountable frameworks. If it is not designed properly, then it can create unintentional harm. It can also produce false information. There are possibilities to generate alternative learning paths in ways that are inconsistent with educational ideals.

This paper tries to research the complicated issues related to the integration of AI in the teaching and learning process. It does this by exploring the fundamental concepts that guide AI adoption, assessing how AI is currently utilized in teaching and learning, and considering the potential risks and ethical challenges that come with its use. Furthermore, it looks into fresh insights about AI's role in formative assessment, personalized learning, and supporting teachers.

11.2 LITERATURE REVIEW

The potential, risks, and consequences of AI in education have garnered significant scholarly interest, especially as it disrupts traditional educational systems worldwide. In the early days, AI was primarily used in education through rule-based tutoring programs and computer-assisted instruction.

DOI: 10.1201/9781003767855-11

But now, thanks to advancements in large-scale machine learning, natural language processing, and generative AI models, educational technologies have evolved to create more interactive, adaptive, and context-sensitive learning environments (U.S. Department of Education, 2023). This literature review encapsulates the role of AI in education, exploring its effects on learning processes, implications for teaching, applications in assessment, and the ethical considerations that shape policy and implementation. It also highlights areas ripe for future research and addresses existing knowledge gaps.

Because AI has interdisciplinary roots in computer science, cognitive psychology, statistics, and human–computer interaction, the literature conceptualizes it in a variety of ways. According to a prevalent perspective, AI can carry out tasks like pattern recognition, language comprehension, and decision-making that have historically required human intelligence. According to educational researchers, AI processes vast datasets to make predictions, categorize data, or produce answers based on statistical correlations rather than replicating actual human cognition (U.S. Department of Education, 2023).

According to Baker and Inventado (2014), a second conceptualization of AI presents it as goal-oriented algorithms that are created to maximize particular goals, like forecasting student performance, suggesting educational activities, or identifying learners who are at risk. However, when models are trained on limited or skewed datasets, algorithmic optimization may clash with more general educational objectives. According to the U.S. Department of Education (2023), algorithmic decision-making may conceal injustices or incorrect data assumptions by appearing more "objective" than it actually is.

Instead of focusing on AI, a third viewpoint stresses intelligence augmentation (IA). This strategy emphasizes how AI can complement human abilities rather than take their place (Holmes et al., 2019). With AI tools as assistants that supplement professional judgment rather than replace it, IA places educators at the center. This emphasizes the value of teacher agency in the adoption of educational technology and is consistent with human-centered design principles.

Finally, AI's reliance on models, that is, mathematical representations of real patterns, also enables us to comprehend this technology. The authors highlight that AI models are merely approximations and thus are influenced by the training data and deployment environments (Clark, 2020). No model can be fully neutral or accurate, which is why the U.S. Department of Education (2023) warns that educators should critically evaluate the findings and maintain oversight.

One of the most frequently cited benefits of integrating AI in education is personalization. For example, adaptive learning enables learning systems and content to be tailored to individual learner profiles. There are indications that such an approach not only increases learning but also enhances student engagement (Pane et al., 2017). As the Department of Education (2023) explains, AI can personalize learning by making recommendations, identifying misconceptions, etc.

Yet, critiques caution against overdependence on adaptive systems. Most of the models retain an overemphasis on narrow and tightly bounded cognitive competencies that are easier to measure, such as procedural mathematics, at the possible expense of broader competencies like creativity, critical thinking, or collaboration (Williamson & Eynon, 2020). Second, personalization risks fragmenting learning experiences, shrinking the opportunities for peer interaction and collective problem-solving.

Intelligent Tutoring Systems, on the other hand, are one of the oldest domains where AI research is most common in education. Meta-analyses show ITS can result in learning gains similar to one-on-one human tutoring, particularly in structured domains such as math and science. Despite these advantages, ITS also has a number of known limitations. ITS tend to be less effective in open-ended domains that involve interpretive reasoning or socio-emotional nuance. In addition, there are concerns about the cultural biases in training datasets and the narrowing of curriculum around things that AI can measure well.

In this regard, AI-based speech-to-text, text-to-speech, translation models, and multimodal supports have significantly enhanced the accessibility for students with disabilities or multilingual background students. Research has shown that it could improve reading comprehension, oral fluency, and content engagement for such diverse learners (Al-Azawei et al., 2017). The Department of Education notice in 2023 conveys an important message: AI can address learning needs not adequately accommodated in traditional classroom structures.

However, experts warn that these "accessibility resources" should be used "advantageously." AI has a high propensity to misunderstand dialects, accents, or expressions unique to a particular culture or ethnicity used in a region. Surprisingly, over-reliance on tech resources has the potential to hinder principled language acquisition.

There is a growing body of research that suggests AI literacy as a subject needs to be incorporated within K-12 curricula. A student needs to have a clear concept of the internal workings of AI systems, the use of data, and the critical assessment of the results provided by algorithms (Long & Magerko, 2020). Algorithmic literacy is becoming a necessity for the future workforce as well as a citizen, as emphasized by the Department of Education (2023).

Nevertheless, teachers feel unprepared to embrace and teach concepts related to AI. The curricular frameworks are also inconsistent. Scholars have emphasized the need to build teachers' skills in the foundations of AI along with a collaborative approach to teaching by embracing computer science concepts and inquiry-based learning (Touretzky et al., 2019).

Research is also showing AI's application in reducing teaching burdens by facilitating the automation of tasks that make teaching easier. According to Holmes et al. (2019), AI is a tool to aid teachers in coming up with easier plans of work, grading tests, creating rubrics, and assisting with a summary of the students. The Department of Education (2023) also adds that AI could be a powerful tool that assists teachers in reflecting on their own practice and making planning easier. With AI, teachers could be able to free up precious time that could be utilized more effectively in personalizing the teaching process, conducting more conferences, and forging close relationships with the students. However, it is also worth noting that there is a possibility of having a few concerns. First, Popenici & Kerr (2017) warn that too much dependence on AI could result in a compromised sense of autonomy for the teachers. The teachers could not fully utilize their own expertise. In addition, there have also been issues around ethics when one considers using AI in teacher evaluation and monitoring. Some studies suggest that, when AI technologies are used in teacher evaluation, this could lead to biases, especially in schools in disadvantaged communities (Benjamin, 2019). Concerns surrounding AI being used in classroom monitoring, specifically face recognition, key pressing, and attention tracking, have also been noted (Knox, 2020). The Department of Education (2023) strongly recommends that teachers must not be evaluated using AI technologies, as this risks being misinterpreted, which could lead to biased outcomes.

Lack of inadequate training and insufficient support from institutions results in lack of confidence in teachers while they use AI (Kessler, 2020). There is a need to create appropriate professional development programs centered on incorporating digital pedagogy, ethics, data literacy, and AI tool assessment (Department of Education, 2023).

AI-based assessment tools can demonstrate how well they are performing in a quick manner. Current studies have shown that learners benefit greatly when they are provided with feedback on their assessments. In particular, studies have shown that when learners are provided with an explanation for the correct answers, they perform better (Shute, 2008). There have been issues over the accuracy of AI-based feedback to the learners.

Evaluation of the writing content by AES systems includes features such as structure, vocabulary, coherence, and grammar. Studies suggest a fair level of correspondence exists between AES systems and human assessment. Nevertheless, it has the potential to save teachers' time and aid students' iterative revision processes. It has also been challenged with creativity, cultural linguistic differences, rhetorical skill, and nonstandard grammar, which sometimes holds students accountable for dialect variations. According to the author, there is a need to highlight the complement of AES systems and human assessment.

A critical perspective on the use of assessments emphasizes that they may include racial, linguistic, and socio-economic biases which are embedded implicitly in the datasets (Williamson,

2019). This can result in some groups being systematically underpredicted in their performance. The Department of Education (2023) has mentioned that AI systems must be continually audited and that data governance will be strategies for risk management. Furthermore, one critical perspective argues that instead of employing single, separate, and high-stakes tests, AI has the capability to create integrated assessment and performance insight systems which are part of daily learning activities (Woolf, 2010). Nevertheless, if the assessment tools are not well-designed, they may cause students to have an excessive cognitive load and hence interrupt learning activities (U.S. Department of Education, 2023).

Studies demonstrate that explainable AI contributes to enhanced user trust, more effective interpretation of results, and also the prevention of the incorrect use of recommendations (Wang et al., 2019).

Findings of the research reveal that there are different levels of access to learning technologies that utilize AI between disadvantaged and privileged learning environments (Bulger, 2020).

The pandemic was behind the increase in the differentials in digital accessibility. For AI in learning, it must be a matter of implementing infrastructure, training, and considering diverse and disparate factors.

11.3 RESEARCH GAPS

The literature points out several key gaps that need attention:

1. There's a lack of long-term, real-world studies on how AI affects various learning environments.
2. Marginalized learners are not adequately represented in training datasets.
3. Inclusion of educators, students, and families in the design process.
4. Lack of strong policy frameworks for the use of AI in education.
5. Teacher preparation programs often overlook the importance of AI literacy.
6. Standards for responsible AI development in educational technologies are still underdeveloped.

The Department of Education (2023) calls for expanded research partnerships to increase funding for AI-in-education research and systematic evaluation of emerging tools.

11.4 THEORETICAL FRAMEWORK

A strong theoretical base assists in explaining the role of AI in effectively supporting teaching and learning. Since AI relates to pedagogy, cognition, ethics, and data science concurrently, different scholars use different frameworks to explain the role of technology in educational environments. This article presents four main theoretical perspectives that are referenced throughout the analysis in this research: (1) Human, Centered AI, (2) Learning Sciences and Modern Learning Theories, (3) Sociotechnical Systems (STS) Theory, and (4) Equity, Centered Frameworks. Together, these theories are in line with the US Department of Education's directive for AI systems that deliver human agency, context, trust, safety, and educational value.

The human-centered AI framework entails that AI should be developed in such a way that it supports human intelligence rather than substitutes it. The framework is chiefly concerned with human–computer interactions. The theory spots out three major areas of focus, i.e., safety, understandability, and human control. Besides that, Shneiderman (2020) explains that this theory is concentrated on human–computer interactions.

In education, the U.S. Department of Education (2023) also holds a similar view: They believe teachers are "in the loop," which means the role of AI is only to assist and not make decisions. In this context, AI supports teachers in analyzing student learning patterns, generating instructional resources, and supplementing feedback. However, teachers retain responsibility for instructional decisions, professional judgment, and ethical oversight. The framework's insistence on inspectable, explainable, and overridable AI also chimes with the department's caution against systems which obscure "how outputs are generated, or which encourage passive acceptance of recommendations."

Thus, HCAI positions educators and learners as empowered actors, ensuring AI operates as a partner rather than an authority.

The application of AI in education is informed by theories of how people learn and how they process information and construct knowledge. There are several learning sciences principles, and the application of AI in education follows these:

The constructivist theory proposed by Piaget, as well as the sociocultural learning theory proposed by Vygotsky, takes into account constructivist learning. The theories recognize the importance of constructivist learning, including effective engagement, collaboration, and co-construction. AI systems that facilitate opportunities for learning, adaptive scaffolding, and effective collaborative learning complement these learning theories (Sawyer, 2014). Intelligent tutoring systems, for instance, use formative feedback that is similar to the zone of proximal development.

However, the Department of Education warns that poorly designed AI narrows instruction to discrete skills or isolated tasks, a conflict with holistic and social approaches to learning.

According to the Cognitive Load Theory, learning is optimized when the instructional materials minimize the unnecessary mental effort of the learners. AI-enabled personalization can support cognitive load regulation by adjusting difficulty, pacing, and modality according to individual needs. For instance, adaptive software can differentiate support for students without overwhelming them with redundant or extraneous information.

Through providing instant feedback, dashboards of students' progress, recommendations for future learning, and opportunities for the development of metacognitive awareness, AI can support Self-Regulated Learning (Panadero, 2017). Learning analytics dashboards and other tools supporting the principles of self-regulated learning have been developed, but the U.S. Department of Education (2023) stresses that these should be very carefully designed to prevent misuse or misinterpretation due to over-surveillance.

Overall, the theories associated with Learning Sciences characterize the use of AI as a means to facilitate deepening and personalizing a student's learning experience, provided that implementation takes into consideration appropriate methodologies.

The STS theory posits that technology and all things relating to it can't be understood alone from the social/cultural/organizational/political environments in which they're implemented (Bijker, 1995). Magisters (teachers) usually work within an education system which is a very human-oriented and contextual system, and thus the effectiveness of AI within that system will be largely reliant on:

- Resources available (from the school)
- Cultural norms
- Teacher training
- Institutional policies
- Community expectations
- Quality of technology infrastructure

According to the U.S. Department of Education (2023), AI tools are going to fail comically if they are introduced without regard to their environment (i.e., localized context). The STS theory demonstrates that AI isn't a neutral entity; it will have an effect based upon how power is distributed, how institutions prioritize issues, and how those same institutions relate with one another.

The following principles should be incorporated when creating an AI framework:

- Actual classroom use involving teachers/students.
- Test repeatedly using a variety of classrooms (i.e., AT LEAST 1 TIME PER WEEK for 12–18 MONTHS, MORE OFTEN is preferred).

- Utilize ethnically approachable materials during the whole expansion procedure.
- Aligns with the mission of the school and the school district.

The STS theory interprets AI as a situation that evolves continuously where humans, organizations, and technology (e.g., hardware/software) have to collaborate/accord a shared vision in order to develop and deliver innovative educational systems.

A last theoretical element is about ethics, fairness, and equity topics that are pivotal in the context of AI integration. Algorithmic fairness frameworks examine how AI may perpetuate or help alleviate inequalities. Studies show that biases are likely to be features of unrepresentative datasets, flawed algorithms, or misaligned institutional goals (Benjamin, 2019). The Department of Education (2023) emphasizes that AI should be checked for discriminatory patterns. It must be excluded from high-stakes decisions that have a direct impact on student discipline, placement, or access to resources. For ethical frameworks of data governance, we have to keep in mind that educational data generally is quite personal and, therefore, it requires very strict protections (Willis et al., 2021).

Openness to share your data and the right to ask for permission before using data are the first characteristics of ethical use of AI. Second, collect as little data as possible from an individual; only the data you need. Third, make sure that anyone who is able to access your data understands what rights they have regarding the use of your data. To determine these goals are essential for responsible and ethical use of AI in education. We must remain vigilant to these discussions. New innovations in AI are consistent and growing so, fair considerations about AI use in education must always be our top priorities. To ensure maximum utilization of AI and lessening disadvantages, we must make sure that AI creates greater equity and fairness for each student, rather than continuing to widen the gap of equity and fairness.

In conclusion, we must take a serious ethical perspective when integrating AI into our education system in order to allow AI to be a positive addition to each student's educational opportunity rather than creating new barriers or reinforcing past barriers to educational equity and fairness.

To lessen the gaps, it is essential to provide resources to schools, i.e., infrastructure, training, or support to adopt new technologies. An equity-focused framework for AI has to be created. It will define various types of opportunities through AI to support a diverse learner population. It should be a resource that recognizes and values students' language and cultural backgrounds and provides inclusive ways to learn together. These same concepts align with the U.S. Department of Education (2023), which encourages using technology as a vehicle for achieving educational equity rather than as a barrier.

11.5 RESEARCH ANALYSIS

AI, being a groundbreaking technology, is radically influencing the entire educational domain, the components of which include, among other things, students, teachers, schools, and the governance framework. Utilizing AI is fraught with numerous problems, which need to be tackled head-on due to its revolutionary character that is changing the face of the education sector. Here, the discussion will be on the most powerful influences of AI, with the discussion focusing on the greatest dents made on students, teachers, assessment, and ethics, along with its influence on the grander concepts of education in general. Hence, the discussion will be weighed against the U.S. Department of Education's (2023) guidelines on the use of AI in the educational field.

AI-powered applications have demonstrated that they are capable of catering to an individual student, customizing a students learning, and changing the material on the go. Next, students' reactions are being employed to make the learning process easier for the students. Years of research have confirmed the fact that these alterations are capable of increasing the students willingness to get involved in the education process, particularly when it is about acquiring new learning skills in math or biology (Pane et al., 2017). In the case when students are not able to succeed in a standard classroom environment, AI-based solutions have opened up the avenue of providing alternative modes of learning.

TABLE 11.1
Ethical and Operational Challenges of Artificial Intelligence in Education

Dimension	Key Issues Identified	Educational Implications	Supporting Sources
Human oversight	Need for human validation and regular bias audits	Risk of over-reliance on automated decision-making	Department of Education (2023)
Assessment practices	Excessive micro-assessments enabled by AI tools	Increased workload; reduced quality of student engagement	Department of Education (2023)
Data privacy	Large-scale collection of student data	Increased risk of surveillance, tracking, and data misuse	Willis et al. (2021)
Equity and access	Unequal access to AI tools and infrastructure	Widening achievement gaps between institutions	Bulger (2020)
Algorithmic bias	Biased training data and predictive models	Discriminatory academic and disciplinary outcomes	Department of Education (2023)
Transparency	Limited stakeholder understanding of AI systems	Erosion of trust among educators and families	Department of Education (2023)

Though personalization has advantages, it also has its drawbacks. Certain models may limit learning experiences for learners by concentrating on quantifiable skills and unintentionally limiting their ability to collaborate, create, brainstorm, or engage in inquiry-based learning (Williamson and Eynon, 2020). The Department of Education (2023) states that adaptive systems must align themselves with holistic learning outcomes instead of providing a pre-set algorithm for learners to follow. When learning systems categorize learners too rigidly, they can reinforce categorization that does not fully represent the ability of that learner.

This Table 11.1 synthesizes Ethical and Operational Challenges of AI in Education. AI-driven accessibility tools such as voice-to-text, text-to-speech, real-time translation, and multimodal content generation provide support to diverse learners, including people with disabilities and people who are multilingual. These tools help to improve participation, reduce barriers due to language, and provide a means of individualized pacing. Several studies show that assistive technologies can improve ease of reading and clarity of expressing ideas (Al-Azawei et al., 2017).

Nevertheless, there is a chance of misinterpreting dialects, accents, or cultural expressions. If such AI systems penalize or give improper feedback for such variations, they may inadvertently support discriminatory practices. Accordingly, it is vital to see that such AI tools for diverse learners depend greatly on cultural sensitivity and evaluation.

One thing teachers usually have a lot of time to deal with includes giving grades, coming up with lesson plans to cater to their students, keeping track of students' attendance, creating examinations or tests, and keeping track of students' performance. Most of this is done using AI techniques and allows teachers to have enough time for other activities, such as teaching and interacting with students (Holmes et al., 2019). According to the Department of Education (2023), teachers have a reduced workload as the immediate benefit of using AI techniques.

AI-enhanced educational planning systems can provide teachers with suggestions for instructional strategies, create rubric development tools, identify resources, and summarize student data. They offer teachers the benefits of virtual assistants and help ease the burden of their workload, leading to a reduction in teacher burnout.

The use of AI in education presents challenges to the professional identity of teachers. An over-reliance on AI-assisted recommendations could erode teacher expertise over time, particularly when AI is perceived to be more "objective" or efficient than human judgment (Popenici & Kerr, 2017). Furthermore, teachers may feel compelled to follow the recommendations generated by algorithms, even when those recommendations do not align with their own contextual knowledge.

AI tools are used by some institutions to track the performance of teachers or the level of student engagement in the classroom, using things like classroom silence, pacing, facial expressions, or the level of digital platform usage. Researchers caution that the use of surveillance-based measures will erode trust, fail to consider the complexity of teaching, and may disproportionately penalize those educators working in high-need settings (Knox, 2020).

Similarly, the Department of Education (2023) has discouraged the use of AI tools in teacher evaluation because of accuracy and discrimination issues, further emphasizing the need to have human-based teacher evaluation systems.

Most teachers indicated that they had limited training to effectively incorporate AI into their classrooms. Without training, teachers may miss out on being able to understand feedback from the AI and use it properly (Kessler, 2020). Such training must also incorporate issues related to digital ethics and test their capabilities against the data.

Moreover, including teachers as co-designers and evaluators helps ensure the validity and contextualization of AI systems and meets human-centered design best practices.

Importantly, AI-based assessment tools comprise an important aid due to their capacity to assess and respond to students' feedback promptly; that is, before a human teacher does, and they aid learners to self-regulate and edit what they have written through identifying patterns and what they don't know and providing individualized practice exercises (Shute, 2008).

Nevertheless, this feedback, although AI-generated, may not at all times be correct. It has, for example, been observed that generative AI models have the tendency to "hallucinate," or make inaccurate explanations, with high degrees of certainty. Automated scoring systems may also misread creative texts, be unable to account for nuances, or misunderstand linguistic differences. Too much emphasis on automated evaluation may also result in formulaic writing and diminish chances for creativity or rhetorical flair (Perelman, 2014).

Furthermore, assessments generated based on AI may contain inherent biases observable in the data used in the past. For example, writing assessments may incline toward standardized academic language, where students using regional dialects and/or multilingual expressions may be at a disadvantage (Williamson, 2019). Similarly, in the observation of "at-risk" students using predictive models, racial and socioeconomic stereotypes may be reinforced.

The Department of Education (2023) puts emphasis on the requirement for human validation, consistent bias audits, and machine learning-based classification restrictions.

While AI makes it easier to assess student performance on a regular basis through micro-assessments, too much testing can cause additional burdens for teachers and learners. Because AI usage typically comes with automated analytics dashboards, teachers may feel pressure to an extent that they must constantly monitor student activity and performance via these dashboards and base learning on the metrics they receive rather than the quality of the engagement. According to the Department of Education, AI should enhance, not increase, the volume of the testing teachers give to their students.

In many cases, AI systems require large volumes of data in order to function effectively, which raises substantial concerns with respect to the privacy of the information being collected about students. Schools indisputably have an obligation to protect their students' sensitive personal data and to limit third-party access to such information. Schools with poorly managed data systems are put at increased risk of students' data being subjected to surveillance, tracking, or commercial exploitation (Willis et al., 2021). The Department of Education (2023) has mandated that any AI systems operating in schools must be in compliance with all applicable privacy laws (i.e., FERPA, IDEA) and must conform to changing privacy policies relating to the protection of students' data.

AI may create greater inequities if different communities do not have the same access to high-quality AIs. For example, poorly funded schools may not have the adequate internet infrastructure, necessary devices, or trained teachers to effectively utilize AI (Bulger, 2020). If wealthier schools implement more sophisticated and innovative AIs while disadvantaged communities do not, the achievement gap between these communities will widen.

TABLE 11.2
Institutional AI Governance Requirements According to the Department of Education

Governance Area	Institutional Responsibility
Decision authority	Clearly define decisions that AI systems may and may not make
Output interpretation	Provide guidance and training for interpreting AI-generated insights
Bias management	Conduct regular bias audits and corrective interventions
Data retention	Specify what data are stored, for what purpose, and for how long
Legal compliance	Ensure adherence to FERPA, IDEA, and evolving privacy regulations

Furthermore, biased training data can result in discriminatory outcomes that will, in particular, affect underrepresented students, especially when predictive analytics are used to decide on course placement or other disciplinary decisions. As such, fair representation within AI designs calls for more representation, culturally responsive content, and participatory development models.

Trust is important in the flawless incorporation of AI. If families and educators do not understand how these systems function, or if schools do not put into place programs without their consent, it can further corrode trust. Communication about the purpose of AI, its limitations, and data policies should be very apparent.

Institutions should develop strategies that spell out:

- What decisions AI is able and not able to make
- how outputs should be interpreted
- how biases are dealt with
- What data are stored and for what amount of time

The Department of Education introduces transparency as one of the major guiding pillars of responsible AI governance.

Table 11.2 shows governance requirements which emphasize transparency, accountability, and ethical stewardship of AI systems in educational institutions.

11.6 CASE STUDIES AND CONCEPTUAL FRAMEWORK

To discover the real-world occurrence of AI in education, it's useful to have examples of the use of AI tools in various situations. Case studies help illustrate the benefits, challenges, and lessons learned by utilizing AI-based instructional systems, teacher helper technologies, and ways to assess student learning through the use of AI applications. Below is a longer description of four different case studies showing how AI is having diverse effects on education: (1) an AI-based adaptive learning system for math, (2) an intelligent tutor platform to help support students who speak a different language, (3) AI-based tools used to automate teacher tasks and decrease personnel needs, and (4) the contentious use of AI-based surveillance and analytic systems in schools. Together, these examples help to illustrate some of the benefits and drawbacks of integrating AI into education and reaffirm the U.S. Department of Education's (2023) focus on implementation of AI that is human-centered, transparent, and equitable.

One of the most researched cases of AI in education involves ALEKS, or assessment and learning in knowledge spaces, a learning platform that has been adopted in many schools and educational systems. ALEKS "uses machine learning algorithms to design individualized learning paths based on each student's knowledge structures and knowledge gaps" (Falmagne et al., 2013). Studies have demonstrated that ALEKS can be highly effective in improving student learning outcomes, particularly for those students who are weak in mathematics. The study of a large number of students revealed that "ALEKS significantly increased these students' passing rates, completion rates, and self, paced mastery of basic math concepts" (Craig et al., 2020).

One of the most effective ways in which AI has been used in education is through the continuous monitoring of the users by ALEKS to deliver learning content on the basis of these observations.

On the other hand, there were some disadvantages to using such a tool. Students learning through ALEKS experienced a lack of social interaction and thus felt like isolates because they had no interactions with co-students and teachers; in some schools, the use of ALEKS resulted in very narrow learning with a sole emphasis on procedural knowledge. Educators also worried that the models on which ALEKS operates were so complex that it was hard to make sense of the recommendations that were given.

These issues are a good match for the Department of Education's (2023) warning that AI tools need to be transparent and remain consistent with the overall learning goals rather than simply making instruction more efficient through the use of algorithms.

AI is creating fantastic opportunities for language learners, especially with the rise of NLP technologies. A great example is ELSA Speak, a mobile app that helps users improve their pronunciation and speaking fluency through AI. It analyzes users' speech, spots any pronunciation errors, and gives feedback based on acoustic models (Nguyen et al., 2019). Literature shows evidences of providing confidence, speaking abilities, and vocabulary to learners while using ELSA and similar AI language tools. As, these tools lessen the pressure of being evaluated by others. It provides ongoing feedback and personalized practice sessions.

According to Luxton-Reilly et al. (2021), these systems are helpful for students to develop their fluency for low-stakes practice opportunities. Nevertheless, this technology presents some limitations. Speech recognition software has difficulty interpreting different accents, dialects, and culturally specific pronunciations, validating and reinforcing dangerous stereotypes about language (Benjamin, 2019). An additional issue identified by teachers is that students sometimes over-relied on the program and were therefore unable to develop authentic conversations/interactions with peers.

The Department of Education (2023) stresses that AI tools for multilingual learners must be culturally responsive and should not replace meaningful human communication or pedagogical strategies.

In this regard, several schools have started incorporating AI-based technologies to support their educators with all their administrative tasks. Gradescope, ClassCraft, and generative AI, among others, are examples of these technological interventions mentioned above. Automated grading of short answer questions, science responses, or writing essays, among others, is slowly taking root in several districts. This, as a result, has seen teachers saving considerably on the time spent on these tasks, such as in a study done by Shermis (2020), where a state in California benefited significantly as a result of incorporating Gradescope to automate STEM tests. The project realized a substantial saving of 60 percent in the time spent by educators on their assessment tasks, as acknowledged by the educators themselves, who mentioned that such ideas have given them an opportunity to prioritize giving individualized feedback or planning interventions for those that are performing poorly. On the other hand, there are several challenges that have since surfaced, with educators being concerned about being biased by generative AI, where an algorithm might focus more on the mechanical soundness of a response, even as it lacks conceptual understanding.

Figure 11.1 presents a conceptual framework that explains how responsible AI governance in education depends on several interconnected pillars, with human oversight at the core. The figure emphasizes that AI should function as a supportive tool rather than an autonomous decision-maker within educational systems.

The Department of Education's (2023) warning regarding AI's capabilities indicates that AI must always be monitored by educators and may not replace judgment calls made by professionals. To maintain the integrity of educator's trust, transparency on how each system provides evaluations is crucial.

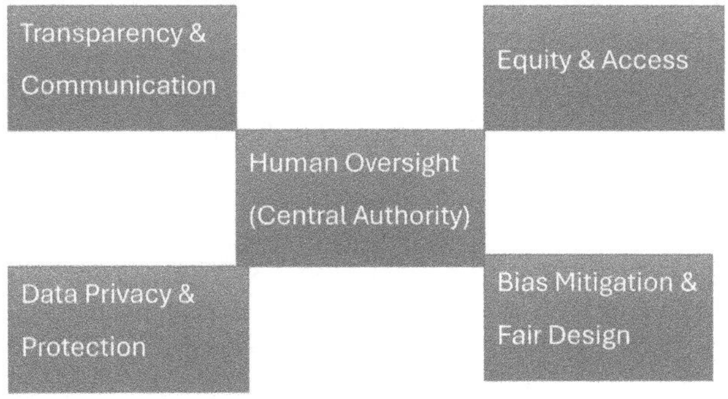

FIGURE 11.1 Pillars of responsible AI governance in education.

More and more districts are piloting AI-based monitoring tools such as facial recognition technology, student activity trackers, and attention analytics during virtual classrooms. Many districts used the breastplate approach with one example of a school district that utilized cameras equipped with AI that detected students' emotional states, inattentiveness, and flagged "behavioral anomalies." Some district leaders may have believed that using AI could simplify the discipline process and report data on students in "real time."

Indeed, some significant problems were discovered by independent evaluations. Students of color were often misidentified by the facial recognition software at significantly higher rates than white students, which is in line with more general findings on algorithmic bias. Students reported feeling constantly watched, which made them anxious reduced their involvement, and increased their mistrust of school officials. The system's labeling of innocuous behaviors as inattentive, such as reading or turning away to think irritated teachers. The system was eventually removed as a result of parents' concerns about data storage, surveillance, and consent. Since these monitoring tools can compromise privacy, dignity, and the emotional climate in classrooms, the Department of Education (2023) has issued a strong warning against AI systems that prioritize surveillance. Numerous examples, including this one, demonstrate the dangers of implementing AI in the absence of candid dialogue, community involvement, contextual awareness, and thorough assessment.

11.7 DISCUSSION

The rising use of AI in school environments has created a wide array of complex opportunities, limitations, and ethical issues. From the literature, case studies, and policy analyses discussed in the previous parts, some major themes can be found that are at the core of understanding what AI means for teaching and learning, equity, and systemic decision-making. This literature review underscores evolving roles of students and teachers, the need for human-centered and equity-focused designs, ethical responsibilities of the institutions, and hence the need for transparency, trust, and continuous assessment of AI use in education. In their totality, these themes point toward the need for deliberate governance and well-contextualized strategies to integrate AI with educational goals that have a real impact.

This is because the AI systems have the tendency to deal with vast amounts of information concerning the students, which is intended to personalize the learning process, give the learner feedback, and even assess performance. The outcomes attained using adaptive systems and learning platforms show that AI can be effective in skill development, although personalization may not inherently foster creativity, socio-emotional development, and collaborative learning (Williamson & Eynon, 2020). The U.S Department of Education (2023) emphasizes that students should not be restricted to undertake their learning by AI systems and are encouraged not to limit their learning potential.

Accordingly, it is necessary to position students as learners who are not simply a recipient of algorithmic instruction but rather an autonomous learner who is aware of the strengths and weaknesses of AI technology. It is necessary to foster AI literacy, as defined by a certain level of awareness about algorithms and data, in order to empower a student with decision-making capacity and critique (Long & Magerko, 2020).

> Research indicates that such an impact is possible: machines could assist teachers, reduce workload, develop resources, and offer insights that inform student performance. Yet, such a consequence is possible only if teachers remain fully in charge. Without humans in the loop, there is a danger that machines could impose their own models of classroom management against teachers' own skills and/or knowledge.
>
> **(Popenici & Kerr, 2017)**

The Department of Education (2023) stresses that it is important for AI systems to be:

- Inspectable: Teachers can understand how results were produced
- Explainability: The recommendation system includes an explanation
- Overridable: Teachers get the option to disagree with or modify the suggestions provided

These principles maintain the professionalism of educators while avoiding the dependency on decision technology devices. They are essential in supporting proper ethical education by ensuring that teachers, not machines, are in control.

Figure 11.2 illustrates the AI-supported assessment and decision-making cycle in education. It shows how student learning activities generate data that are analyzed by AI systems and presented through analytics dashboards. These insights are then reviewed by teachers, whose professional judgment guides instructional decision-making. The figure emphasizes that AI functions as a support tool, while educators retain final authority over pedagogical decisions.

AI can support students who are multilingual, have disabilities, or require tailored support, according to several studies. However, AI tools should be created with a variety of users in mind in order to work effectively. Training datasets for AI tools frequently have elements of bias against people from groups that are underrepresented (for example, dialects and cultural expressions not typically represented), and they also have criteria for proficiency that often favor students who are within the "norm" set by the algorithms created by developers (Benjamin, 2019). To create an equitable AI tool, the following are required:

- The creation of datasets that are racially and culturally representative
- A design process that is inclusive and incorporates student and educator input
- An ongoing audit of bias in the algorithm
- The adaptation of data based on the characteristics of the diverse learning environment

FIGURE 11.2 AI-Supported assessment and decision-making cycle.

The Department of Education (2023) thus advises those who develop the new technologies to "center inclusion and fairness," but most importantly, "must actively work towards eliminating existing inequities."

Further, predictive analytics also carries its own risks, as a "flagging" system relying on past patterns may even emphasize existing stereotypes based on race, language, disability, or socio-economic status. "There is even a concern, as a result, that predictive analytics tools, through their effect on teachers' perceptions of students, might have a subsequent impact on relations with those students" (Williamson, 2019).

Consequently, AI must be made to function as a support, but not as a predictor of a student's potential. Institutions need to develop potent policies that dictate how AI collects, stores, seeks consent, and accesses its information. Schools have sensitive customer information that should be handled with care. AI usually needs a lot of information to perform well in most cases. These systems can violate student privacy without proper guidance (Willis et al., 2021).

Insufficient teacher preparation consistently appears as a concern in many studies. For example, too many educators have reported feeling unprepared to assess the quality of AI tools, make sense of algorithmic outputs, or catch biases, according to Kessler (2020). Successful integration demands ongoing, job-embedded professional learning on topics that range from digital ethics and data literacy to instructional integration to codesign strategies.

Additionally, teacher involvement in the process of evaluation and tool development reinforces the depth of contextual relevance and minimizes the prospect of technology imposition from the top down. Schools should indicate clearly which AI technologies are inappropriate, such as those for surveillance purposes, emotion detection, biometric monitoring, or automatic scoring of behavior. Case studies' evidence reveals that such tools break down trust and cause a misinterpretation of student behavior, additionally, they have a negative impact on marginalized groups more than others (Knox, 2020). The Department of Education (2023) has issued a strong warning against such practices, stating that they are not in line with ethical, educational, and developmental principles.

Trust is the fundamental element for the acceptance of AI, and trust can only be built when there is a clear understanding of why systems work as they do and how they make decisions. When algorithms are hidden, they decrease people's trust in AI systems and can result in the systems being used wrongly or unethically. Therefore, it is very important to make sure that explainable AI models are created ethically so that they can be the basis for the responsible use of these technologies (Wang et al., 2019).

In order to come up with effective strategies for implementing AI in classrooms, there should be a collaboration between teachers, software developers, school leadership, policymakers, and families. When we work together to design the tools and resources for classrooms, we can develop solutions that will address the specific challenges faced by teachers and the diverse needs of learners. Moreover, the sharing of responsibility enables the introduction of new technologies to be balanced with the understanding that educators have to base their teaching on their own knowledge and experience rather than solely on the use of technological solutions.

11.8 CONCLUSION

In just a few years, AI will have changed how we educate students and improve the process. The innovations provided by AI are beyond what was thought possible only ten years ago, as adaptive learning technologies, intelligent tutoring systems, and automated assessment systems appear to be very promising for creating more efficient methods of teaching, delivering more personalized learning opportunities to all types of learners, and leveraging technology to assist teachers in succeeding with all students. However, as shown in this research study, the advantages of AI are not given automatically or equally. When AI is integrated into educational systems, there are complex issues of transparency, privacy, fairness, identity of professionals, and the equity of all individuals involved. Due to these new realities with the advent of advanced educational technologies such as

AI, it is important and prudent to use a measured approach when deciding to implement or use AI in schools.

The takeaway from various literature, research, and policy studies is that AI can significantly help in decision-making but should not be made to replace human judgment at any time.

Excellent teaching is a human, interpersonal, and deeply understanding, oriented practice that is especially geared to the learners' educational, personal, and cultural backgrounds and their needs and desires. Although the technology may be great in gathering huge data and spotting patterns, it is still incapable of learning anything from a good teacher.

The U. S. Department of Education (2023), in a policy brief, identifies this difference and points out that inspectability, explainability, and overridability are key features of evaluating AI systems that should allow teachers to have complete decision-making authority.

The findings of the research emphasize equity and equality in developing and using AI. In the lack of ethical usage of AI, it may result in social inequality for multilingual learners, especially abled, and other marginalized learners. And that would be due to biased training data, unclear model designs, and inaccessible tools. That is why there is a great need for the use of diversified datasets, taking culturally responsive design practices, conducting regular bias audits, and making sure that educators and students who best know the realities of the classroom contribute to actively shaping AI in education.

In addition to the issue of governance and ethics, schools and school districts must differentiate their positions on the issues of data privacy and transparency, as well as third-party entities, in detail in the form of a document. Besides, the policies must spell out the use of AI and based analytics that can be deemed good. The teachers must have professional development that ensures they are well-trained in the use of AI in a critical manner and with full faith in themselves. Ultimately, the major thing that comes to mind is the fact that we must not use AI as a system of surveillance and behavioral scoring since it goes against teaching values, and such use constitutes dehumanizing the students.

Ultimately, we need to ensure that we're crafting the future of AI in ed tech with intention and human touch. The potential of AI in transforming education in positive ways is simply huge and welcomed in line with current learning theory. It is important to highlight that in crafting the future of ed tech with AI, we need to ensure that we're not crafting something that detracts from our goals. When looking to the future of AI in ed tech, gazing into our crystal balls, we need to ensure that all who participate in ensuring the success of education, including ed tech and policy research, need to be making concerted efforts to work together.

REFERENCES

Al-Azawei, A., Serenelli, F., & Lundqvist, K. (2017). Universal Design for Learning (UDL): A content analysis of peer-reviewed journal papers from 2012 to 2015. *Journal of the Scholarship of Teaching and Learning*, *17*(3), 1–23. https://scholarworks.iu.edu/journals/index.php/josotl/article/view/19295

Baker, R. S., & Inventado, P. S. (2014). Educational data mining and learning analytics. In J. A. Larusson & B. White (Eds.), *Learning analytics: From research to practice* (pp. 61–75). Springer.

Benjamin, R. (2019). *Race after technology: Abolitionist tools for the New Jim Code*. Polity Press.

Bijker, W. E. (1995). *Of bicycles, bakelites, and bulbs: Toward a theory of sociotechnical change*. MIT Press.

Bulger, M. (2020). The digital divide and educational equity: A look at the challenges. *Journal of Educational Technology & Society*, *23*(4), 1–10.

Clark, A. (2020). Model complexity and the limits of AI prediction. *AI & Society*, *35*(4), 987–996.

Craig, S. D., Chi, M., & VanLehn, K. (2020). Improving learning outcomes with intelligent tutoring systems: Evidence from meta-analysis. *Review of Educational Research*, *90*(4), 499–537.

Falmagne, J.-C., Cosyn, E., Doignon, J.-P., & Thiéry, N. (2013). *Knowledge spaces: Applications in education*. Springer.

Holmes, W., Bialik, M., & Fadel, C. (2019). *Artificial intelligence in education: Promises and implications for teaching and learning*. Center for Curriculum Redesign.

Kessler, G. (2020). Preparing teachers for AI-enriched classrooms. *Language Learning & Technology, 24*(3), 1–10.
Knox, J. (2020). Machine behaviourism: Future visions of 'learnification' and 'credentialization' in AI-based education. *Learning, Media and Technology, 45*(1), 31–45. https://doi.org/10.1080/17439884.2019.1623251
Long, D., & Magerko, B. (2020). What is AI literacy? *Proceedings of the 2020 CHI Conference on Human Factors in Computing Systems*, 1–16.
Luxton-Reilly, A., Alizadeh, A., & Seagram, M. (2021). Using AI-based tools to support language learners: Benefits and limitations. *Computers & Education, 168*, 104195.
Nguyen, H. T., Nguyen, A., & Vu, T. (2019). Automated pronunciation training for language learners. *International Journal of Artificial Intelligence in Education, 29*(2), 231–254.
Pane, J. F., Steiner, E. D., Baird, M. D., & Hamilton, L. S. (2017). Informing progress: Insights on personalized learning implementation and effects. *RAND Corporation.*
Panadero, E. (2017). A review of self-regulated learning: Six models and four directions for research. *Frontiers in Psychology, 8*, 422. https://doi.org/10.3389/fpsyg.2017.00422
Perelman, L. (2014). When "the state of the art" is counting words. In B. A. Slater & P. Y. Mosenthal (Eds.), *The mismeasure of writing: The academy's response to automated essay scoring in holistic assessment* (pp. 128–143). Hampton Press.
Popenici, S. A. D., & Kerr, S. (2017). Exploring the impact of artificial intelligence on teaching and learning in higher education. *Research and Practice in Technology Enhanced Learning, 12*(1), 1–13. https://doi.org/10.1186/s41039-017-0062-8
Sawyer, R. K. (2014). *The Cambridge handbook of the learning sciences* (2nd ed.). Cambridge University Press.
Shermis, M. (2020). Contrasting state-of-the-art automated scoring of essays: Pearson and ETS. *Computers & Education, 150*, 103855.
Shneiderman, B. (2020). Human-centered AI. *Issues in Science and Technology, 36*(2), 1–10.
Shute, V. (2008). Focus on formative feedback. *Review of Educational Research, 78*(1), 153–189.
Touretzky, D. S., Gardner-McCune, C., Martin, F., & Seehorn, D. (2019). Envisioning AI for K–12: What should students learn? *ACM Inroads, 10*(4), 34–40.
U.S. Department of Education, Office of Educational Technology. (2023). *Artificial intelligence and the future of teaching and learning: Insights and recommendations.*
Wang, D., Yang, Q., Abdul, A., & Lim, B. Y. (2019). Designing theory-driven user-centric explainable AI. *Proceedings of the 2019 CHI Conference on Human Factors in Computing Systems*, 1–15.
Williamson, B. (2019). Policy networks, performance metrics, and algorithmic governance in education. *International Journal of Qualitative Studies in Education, 32*(4), 406–425.
Williamson, B., & Eynon, R. (2020). Algorithmic imaginaries of education: A future-oriented lens. *Learning, Media and Technology, 45*(1), 1–15.
Willis, J., Slade, S., & Miller, R. (2021). Ethical data practices in schools. *Phi Delta Kappan, 102*(6), 20–25.
Woolf, B. P. (2010). *Building intelligent interactive tutors.* Morgan Kaufmann.

12 Environmental Sustainability of IoT and AI Deployments in Education

Pathways, Challenges, and Green Solutions

Ajay Kumar, Yahya Fikri, Sanjeev Kumar, Kuldeep Singh, and Redouane Kaiss

12.1 INTRODUCTION

The entire education sector in the world is shifting irreversibly and unparalleled because of the rapidly changing digital technologies. A shift toward the incorporation of information and communication technology (ICT), big data, cloud computing, and intelligent automation is a trend that has been witnessed in the past two decades to change the nature of education delivery, access, and management (Bolón-Canedo et al., 2024; Fikri et al., 2025). These technological changes include the Internet of Things (IoT) and Artificial Intelligence (AI), which have become the staples of the future of teaching, learning, and institutional governance. These technologies are not the peripheral gadgets anymore, but are now part of the procedure and approach of the contemporary educational institutions (Mallo et al., 2024). The systems will establish added efficacy in teaching, learning customization, elevated student retention and effectiveness of the institutions (Omar et al., n.d.). The combination of IoT and AI has helped promote data utilization in decision-making, gathering real-time data about learning environments, as well as automation of administrative activities, thus, transforming the traditional boundaries of the education system (Ullah et al., 2024).

Existence of IoT and AI in the educational sphere has further intensified and speeded up the rapid evolution due to globalization, convergence in technologies, and post-pandemic development of free and hybrid model educational activities (Das & Chandra, 2023). This comparative transformation brought forth the opportunities and threats of online education systems. results in the state whereby learning institutions are found operating on data-intensive digital ecosystems, where the learning processes, student behavior, and administrative functions are regularly being monitored, recorded, analyzed, and maximized (Eshbayev et al., 2024). However, despite the increased concern about the problem of climate change and environmental degradation expressed by the growing number of people worldwide, the aspects of sustainability are typically overlooked in the planning and the digital education policy making (Ibrahim et al., 2024).

The education sector and the establishment of IoT and AI can be attributed to a number of interdependent technological, economic and institutional factors. First, the declining cost of sensors, connection, cloud service, and AI tools made the increasingly advanced digital infrastructure possible to more schools and universities, including those located in the less developed economies (Chigbu & Makapela, 2025). The low price of smart devices and cloud services that are subscription-based have enabled entry barriers to allow scale-based implementation of both IoT-enabled and AI-enabled solutions in institutions (Mohsen, 2024). The traditional one size fits all teaching model

 DOI: 10.1201/9781003767855-12

is slowly being perceived as not being adequate to accommodate the diverse needs of learners. Adaptive learning patterns can be implemented with the help of intelligent tutoring systems and AI-based learning analytics based on the analysis of the trends in the learner behavior, performance, and engagement (Wu et al., 2024). Schools are in a very competitive environment, in which performance measures, accreditation criteria and world ranking are the variables that influence the strategy (Goel et al., 2024).

12.1.1 Digital Transformation and Environmental Effects

Even though the advantages of digital transformation in the educational field cannot be overvalued, the effects of widespread use of IoT and AI on the environment are more apparent (Nwaigbo et al., 2025). The data processing, data storage and transmission energy expenditure has increased, as the online learning platform, video streaming, and AI-based analytics multiply by folds (Ogbanga et al., 2025).

The main component of AI-enabled education systems, the data centers, fall into the group of the most energy-intensive components of digital infrastructure (Kataria et al., 2024). This energy requirement has risen with the growing adoption of cloud-based learning management system, AI marking system and virtual laboratories (Mahmood et al., 2024). Institutions of learning tend to upgrade smart boards, servers, tablets, sensors, and networking gadgets so as to keep up with the technological change. E-waste dumping and recycling is the biggest ecological and health-related hazard, particularly in the rural regions where several individuals engage in informal recycling practices (Yigitcanlar et al., 2020; Fikri et al., 2025).

The environmental effect of the production of the device, the extraction of the materials, the consumption of electricity and the storage of information and the character of the disposal confirm the popular opinion that digitalization is inherently sustainable (Keskar, 2022). Even though digital education has the possibility of reducing the quantities of paper and physical travel, such benefits may be succeeded by the ecological footprint of the digital infrastructure, provided that the sustainability is not directly addressed (Mak et al., 2022).

The responsible consumption, climate action, and sustainable infrastructure are the aspects of sustainable development that are outlined but are not usually regarded as the element of the educational technology strategy (Yigitcanlar et al., 2020). The chapter addresses the gap in the new discussion of the subject of green educational technology and responsible digital transformation through reviewing the existing studies, institutional practices, and sustainability frameworks.

12.2 REVIEW OF LITERATURE

The IoT and AI literature in education are also on the rise over the past decade. To a larger part, this research is though anchored on pedagogical innovation, learning outcomes and institutional efficacy but not on environmental sustainability.

12.2.1 Introduction: Positioning of IoT and AI in Education Studies

Detailed investigations of educational technology in the past two decades have ceased small, tool-based research, in favor of multi-system investigations of digitally mediated learning ecologies. IoT and AI represent a new level of these ecologies: IoT can provide continuous and high-resolution environment and behavior data, whereas AI can provide offer inference, automation, and personalization services (Rawat & Yadav, 2025). Scholarly Pedagogically, scholars being concerned with the possibilities of improved learner diagnosis, adaptive scaffolding (e.g., intelligent tutoring, automated formative assessment) and administrative efficiency. The socio-technical issues raised by authors include governance, privacy and equity (data ownership, algorithmic bias). However, a new subrident has started to believe that sustainability of the environment is a first order variable in the design and evaluation of such systems – something that this chapter follows (Fan et al., 2023). Such lenses (life-cycle assessment, circular economy, energy-conscious design) that are being applied in education technologies can be offered by theoretical framing Sustainable ICT and socio-technical systems literature.

12.2.2 IoT Devices, Smart Classroom and Material Dimension

Interactive whiteboard, environmental sensors (temperature, CO_2, occupancy), RFID access/attendance, tablets and mobile devices, cameras and an educational facility-management IoT (lighting, HVAC controls) are examples of IoT devices in education. It has been demonstrated in the literature on pedagogy: better engagement, context-oriented learning, and better campus management (e.g., energy optimization with smart HVAC in relation to occupancy sensing). Such devices are however contingent in materiality. Their manufacture needs metals and plastics, the electronics used in them requires rare earths and critical minerals; their networks and software need continuous maintenance and regular updates as standards evolve (Dhayal et al., 2025). Two tensions have been determined as a result of empirical work. The presence of e-waste monitored around the globe indicates an increase in more e-waste which must be considered by educational purchases (see in detail in Section 2.4). Recent technical review also suggests that education IoT should have design-to-last-long and modular hardware to reduce replacement rate and refurbishment.

12.2.3 AI Model and Computational Footprint Learning Analytics

Educational learning analytics and AI models lie between basic classification/regression models which predict dropouts and deep-learning models which predict natural-language feedback and scoring academic essays. Several studies are already showing that machine learning model selection and pattern of deployment has a large quantitative effect on energy use: simple models (logistic regression, decision trees) are usually inexpensive in terms of runtime energy (as well as fairly expensive to train); and can be run either locally or on low-power servers (Allioui & Mourdi, 2023). Deep models and large language models (LLMs) have orders of magnitude higher energy requirements to train and (some deployments) inference energy requirements. Even though scale is not critical to most educational implementations of AI, the tendency to resort to cloud-based, ever-fine-tuned models is increasing the amount of energy demanded by learning analytics systems.

12.2.4 Cloud Dependence and the Macro Energy Picture Data Centers

Data centers are one of the most important strands when linking the decisions made at the education level with impacts on the planet. LMS hosting, video streaming, storing student records and AI inference pipelines are some examples of cloud-based services that are the foundation of numerous educational platforms (Kumar et al., 2024). The International Energy Agency (IEA) estimates that data-center power usage may rise by 2030 to around 945 TWh in a central case, approximately 3 percent of world electricity consumption in a central case in case this trend persists. By itself, systematic review and national studies project an estimate of data-center consumption of between hundreds of TWh at present. These statistics are important in education, as even a small increase in online lecture streaming, AI-supported assessment and high-frequency analytics of millions of learners around the world is relevant (Sanusi et al., 2023).

There are two more aspects that matter. First, there are issues related to the source of electricity: a data center using grid electricity generated using coal is much more intensively carbon-intensive than one using renewables or grid mixes with high renewable content. Second, non-energy water and material footprint is also being seen as co-drivers of environmental impact and is being given additional focus in the practical sustainability evaluation of IT infrastructure (Sanusi et al., 2023; Fikri & Rhalma, 2024).

TABLE 12.1
Environmental Impacts and Sustainability Responses of IoT and AI in Education

Component	IoT/AI Application in Education	Key Environmental Impact	Sustainability Challenge	Proposed Green Solution
Smart classrooms	Sensors, smart boards, surveillance systems	High electricity consumption	Continuous power demand	Energy-efficient sensors, smart scheduling
AI learning analytics	Predictive models, automated assessment	High computational energy use	Carbon-intensive cloud dependence	Model efficiency, edge computing
Cloud infrastructure	LMS, video streaming, data storage	Rising carbon footprint of data centers	Fossil-fuel-based electricity	Green data centers, renewable energy
Digital devices	Tablets, servers, networking equipment	E-waste generation	Short lifecycle and obsolescence	Circular procurement, refurbishment
Campus management IoT	HVAC, lighting, access systems	Material and energy intensity	Poor lifecycle governance	Modular design, lifecycle management

12.2.5 Synthesis and Research Gaps

These three areas of intersection exist in the literature. First, the significance of both IoT and AI has been demonstrated with pedagogical and operational benefits, and their application in the education field will continue to increase. Second, the environmental cost of these systems, especially the energy use of the data-centers, the lifecycle emissions of the devices, and the e-waste is not minimal and it is growing. Third, there is limited research on participation and application of technical and governance solutions (edge computing, model efficiency, green procurement, circular economy approaches) and their application in educational setting is unevenly spread across available empirical studies. The research gaps that still warrant the empirical interest of this chapter include: definite carbon- and energy-benefits of accounting of typical educational workflows (e.g., a semester-long course of blended course) and case-based analysis of the adoption of institutional circular procurement and policy studies of the relation between national e-waste and green-ICT policies and higher-education procurement practice.

Table 12.1 contains synthesized information on the major IoT and AI applications applied in educational institutions and associated environmental implications. It provides a systematic connection between technological elements like smart classes, AI-based learning analytics, cloud computing, digital devices and campus management systems with its main environmental effects, which include energy use, carbon emissions and e-waste production. The table also lists the key challenges of sustainability, including 24/7 power supply, limited lifecycle of devices, and data centers that rely on fossil fuels. Notably, it puts emphasis on viable green solutions, such as energy-efficient sensors, model optimization, green data centers, circular procurement and lifecycle management. In general, the table appears to be a brief analytical framework that fills the gap between digital transformation and sustainability efforts in education.

12.3 RESEARCH METHODOLOGY

The chosen research methodology in this chapter is the qualitative one that will be applied to discover the environmental sustainability questions concerning the application of IoT and AI technologies in higher education institutions in a well-organized manner. The problems of sustainability in digitally transformed education systems have recently been examined, and a qualitative approach is particularly suitable to acquire in-depth information, institutional practice, and perspectives of the involved stakeholders, which are often overlooked during the quantitative solution. The methodology design is geared toward the realization of analytical rigor, transparency and credibility yet needs to be flexible so as to capture the complexity of socio-technical and environmental problems.

This study has a qualitative and exploratory research design, which is a descriptive and interpretive research design. The exploratory nature of the design is justified by the fact that the environmental sustainability of the IoT/AI-based education is a largely unexplored field, especially in the institutional practices and decision-making. Rather than attempting to test pre-established hypotheses. This combination allows the research to go an additional step further than the superficial accounts and argue with the internalizations, limitations, and interests of institutional action.

The participants were selected through a purposive sampling design because the participants are the most immediate and engaged in terms of objectives of the research. In qualitative studies the understanding object of interest is sought and not statistical extrapolation and purposive sampling is appropriate. Faculty members who are active in the use of digital platforms, tools of learning analytics and AI-based systems of teaching or assessment participated in the teaching practices. The study employed several qualitative methods of data collection in an attempt to gather all the information, and this employed both primary and secondary data. This triangulation strategy enhanced validity and richness of the findings.

12.4 RESULTS AND THEMATIC ANALYSIS

This section presents the findings, which have been reached according to the qualitative analysis of the data of the interview and the institutional documents. Using a systematic thematic analysis approach and the help of the MAXQDA tool, the researchers discovered the list of themes that are reiterative and reflect the perceptions, management, and responses of educational institutions to the emerging environmental sustainability issues that are caused by the introduction of IoT and AI technologies. These findings are divided into large themes and sub-themes and each of them explains a specific component of the sustainability-digitalization nexus in higher education. The study reveals a high degree of the disalignment between the objectives of digital transformation and the environmental sustainability practice despite the rise of awareness of the institutional stakeholders.

THEME 12.1: DIGITIZING WITH LIMITED SUSTAINABILITY INTEGRATION TECHNOLOGICAL CHANGE IS DRIVING DIGITAL TRANSFORMATION

One of the most powerful themes that were brought out in the data were the technology-centered nature of digital transformation in learning institutions. The objectives of implementing the IoT and AI were also the concern of the participants who continuously emphasized that the major objectives in the decisions to implement the modernization, competitiveness, student contentment, accreditation conditions, and operational effectiveness. Environmental sustainability was not considered as one of the basic design requirements and, in fact, was assumed to be an indirect or optional problem. It was observed by the administrators of academic institutions that there is a call to adopt the latest technologies in order to enhance an image of the given institution and the international rankings. The members of the faculty were concerned with the effectiveness of the pedagogy and interaction of students offered by AI-based platforms. IT managers focused on the performance of the systems, scalability, and security.

THEME 12.2: SOARING ENERGY CONSUMPTION AND INVISIBLE CARBON FOOTPRINTS

The second most important issue associated with the high energy consumption in connection with the introduction of IoT and AI. All categories of respondents acknowledged that smart classes, AI-based applications, and cloud-based solutions were a significant factor in electricity consumption without much active monitoring or quantification of the impact in most of the institutions. IT managers defined incremental power requirements in servers, network equipment, biometrics and surveillance equipment and sensor networks. Faculty members have said that they are expanding the high-bandwidth projects such as video streaming, virtual labs, and AI-powered evaluation. Such energy demands however were not usually translated into measurable carbon footprints on institutional level. The invisibility of digital emissions was found to be one of the most significant ones.

THEME 12.3: E-WASTE CREATION AND LACK OF LIFECYCLE MANAGEMENT

The third general theme has to do with the generation of electronic waste (e-waste) and the inappropriate lifecycle management approaches. Almost all the institutions that were part of the study asserted that they constantly updated their digital tools including smart boards, servers, networking devices, tablets, and IoT sensors. Compatibility issues with software, the maintenance cycle of the vendor, and perceived obsolescence were factors involved with such upgrades. In spite of the fact that the quantity of the devices turned over was high, the e-waste management systems were not formal. The respondents stated that old devices were kept too long, sold in the informal manner or deposited in general waste.

THEME 12.4: CONSCIOUSNESS OF SUSTAINABILITY DILEMMAS AND INADEQUATE INSTITUTIONAL RESPONSE

The level of conceptual awareness and low level of concrete action was an interesting and even contradictory observation. Many of the respondents had certain concerns associated with climate change, energy use and degradation of the environment. The faculty and the administration frequently confessed to possessing the concealed expenses of digital technology on the environment. However, such a perception did not necessarily lead to institutional action or policy reform. The participants have stated several barriers to acting and these include:

- Budget limitations and cost focus.
- Low technical savvy in green ICT.
- Lack of regulation requirements.
- Disjointed administrative systems.

Sustainability had often been described as token, rather than radical, limited to such one-off campaigns as lighting campaigns (energy-saving), paper campaigns (reduction). The digital infrastructure in its turn was largely left out of the sustainability agendas. The implication of this observation is that sustainability cannot be part of digital decision-making, no matter how much awareness is available, but there are no institutional structures, incentives, and mechanisms of accountability that will make sustainability a fundamental aspect of digital decision-making.

12.5 DISCUSSION OF RESULTS

In this section, the author targets to elucidate the case study in the context of extant literature, theoretical approaches and contemporary discussions of sustainable digital transformation of education. Despite the presentation of the above section which is the presentation of the themes according to the qualitative analysis, the discussion here contextualizes the empirical themes, elaborates on its overall applicability and the way the research will contribute to the body of knowledge regarding the environmental sustainability of IoT and AI implementations in learning institutions. Overall, the evidence shows that technological advancement and environmental sustainability is imbalanced in terms of structure. IoT and AI technologies to enhance efficiency, personalization, and competitiveness are highly popular, however, sustainability is either marginal, fragmentary, or reactive.

12.5.1 Digitalization as Technologically Focused and Sidelining Sustainability

The most significant finding of the chapter is that digital transformation in education is, to a large extent, technology-oriented, yet not sustainability-oriented. This is a continuation of research works that have indicated that the majority of learning institutions are more inclined to embrace digital technologies with a focus on innovation, performance indicators and market positioning as opposed to environmental policies.

12.5.2 The Energy Consumption and the Unspoken Carbon Footprint of Digital Education

These outcomes connected with the increasing energy consumption and the implicit carbon footprints show one of the most critical inconsistencies in the contemporary educational systems. Though the concept of digital education is promoted as an environment-friendly one, since it leads to the reduction of paper use and travel, the study reveals that the energy and carbon costs of it are not properly understood and can barely be measured in the institutional level. This observation is equivalent to the literature on digital sustainability that identifies digital emissions as an outsourced or invisible one.

12.5.3 E-Waste, Obsolescence and Linear Technology Consumption

Poor lifecycle management and e-waste production is one of the themes that signify a linear concept of technology adoption that dominates the educational procurement setting. Devices are bought, consumed in relatively low quantities and either discarded or replaced without any long-term reuse or recycling strategy. This trend is indicative of the tendencies in the world toward the use of electronic but it is particularly scary in connection with the educational process due to the scale and the pace of the devices introduction. The findings are rather audible as far as the literature on the circular economy is concerned, and the linear take-make-Dispose models are unsustainable in terms of environmental considerations.

12.6 IMPLICATIONS OF THE STUDY

The above evidence and discussion suggest that the problem of environmental sustainability concerning the implementation of the IoT and AI in education is not necessarily technical, and has a solid institutional foundation on the basis of institutional decisions and governance structures and policy environments. This section will include the practical implications on the basis of empirical findings to the practical managerial and policy implications and will serve as a guide to the education leaders, the administrators, the technology managers and the policymakers who are determined to make digital transformation a responsible aspect environmentally. Findings indicate that there is no sustainability in the digital vision statements and ICT roadmaps, which implies that sustainability will not have any impact on operational decision making regarding the implementation of IoT and AI. Managers are thus advised to ensure that the sustainability objectives are explicitly mentioned in digital transformation strategies. It can be argued based on the study results that the digitalization policies of education need to be more consistent with the global sustainability policies such as the United Nations Sustainable Development Goals. In particular, the responsible consumption and production, climate action, and sustainable infrastructure goals have a direct applicability to the IoT-based education system incorporating AI. Accountability and adoption of best practices can be influenced by policy makers by directly relating the international commitment to sustainability to digital education activities. Such alignment also makes international benchmarking and knowledge exchange very easy.

Figure 12.1 represents the sustainability trajectory of IoT and AI implementation in education through the how technology integration is moving toward environmentally friendly digital ecosystems. It starts with the implementation of IoT and AI applications (smart classroom, learning

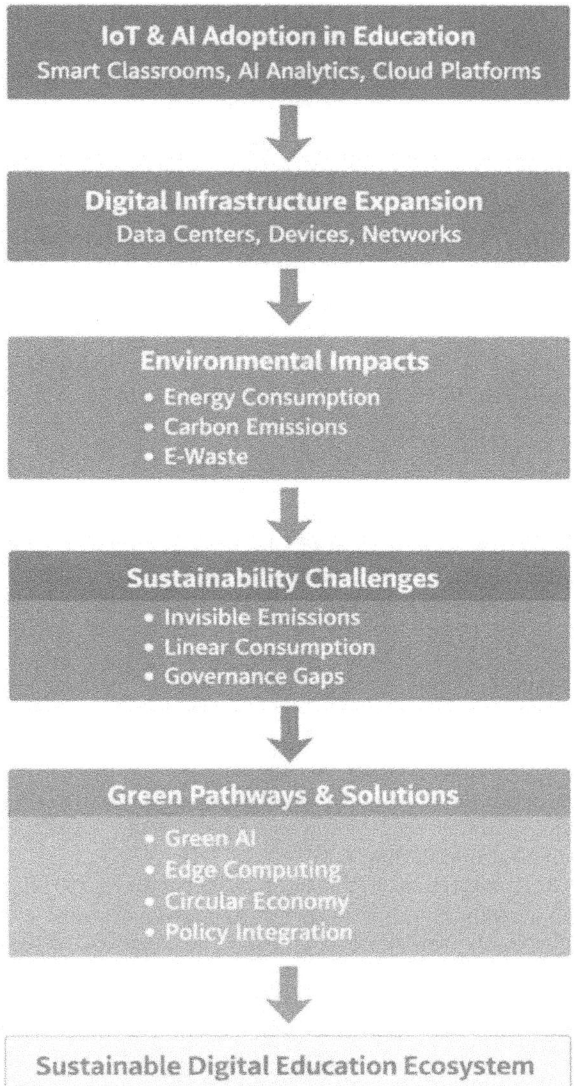

FIGURE 12.1 Sustainability pathways for IoT and AI-enabled education systems.

analytics, and cloud platform) that result in the growth of digital infrastructure (data center, network devices, and network). The growth has environmental effects on the growth of energy usage, carbon emissions, and generation of e-waste. The figure then points out major sustainability issues which are invisible emissions, linear consumption patterns and missing governance. Finally, it introduces green directions, including Green AI, edge computing, circular economy activities, and policy integration, which all make it possible to create a sustainable digital education ecosystem.

12.7 CONCLUSION

This chapter sought to critically examine the environmental sustainability concern of the applications of IoT and AI technologies within learning institutions. According to the qualitative, exploratory research design and thematic analysis of the views of the stakeholders, the study provides empirical evidence related to the existing conceptualization, implementation, and control of the digital transformation in education, in this case in the environment of environmental responsibility. The findings confirm the fact that IoT and AI are already entrenched in the new system of education and transform the process of teaching and learning process, learning environment, and institutional governance. Smart classrooms, learning analytics on the basis of AI and cloud-based systems, as well as automated administrative systems have already become not only an experimental innovation but also a component of digital education ecosystem. Such technologies are quite helpful in terms of efficiency, personalization, access, and data-driven decision-making. However, the study shows the high levels of environmental costs in relation to these benefits that are not seeable and are not well regulated. Overall, this chapter contributes to the literature since it concludes that the sustainability challenges associated with IoT and AI in education are not technical, but institutional, governance, and policy in nature. In order to challenge them, it is significant to shift the conceptualization of digital transformation, yet not to emphasize efficiency and innovation but more survival-oriented concept of educational progress.

REFERENCES

Allioui, H., & Mourdi, Y. (2023). Exploring the full potentials of IoT for better financial growth and stability: A comprehensive survey. *Sensors*, *23*(19), 8015.

Bolón-Canedo, V., Morán-Fernández, L., Cancela, B., & Alonso-Betanzos, A. (2024). A review of green artificial intelligence: Towards a more sustainable future. *Neurocomputing*, *599*, 128096.

Chigbu, B. I., & Makapela, S. L. (2025). AI in education, sustainability, and the future of work: An integrative review of industry 5.0, education 5.0, and work 5.0. *Journal of Open Innovation: Technology, Market, and Complexity*, 100645.

Das, K. P., & Chandra, J. (2023). A survey on artificial intelligence for reducing the climate footprint in healthcare. *Energy Nexus*, *9*, 100167.

Dhayal, K. S., Giri, A. K., Agrawal, R., Agrawal, S., Samadhiya, A., & Kumar, A. (2025). Do the innovative technological advancements foster the green transition pathways for industry 5.0? A perspective toward carbon neutrality. *Benchmarking: An International Journal*. https://www.emerald.com/insight/content/doi/10.1108/BIJ-04-2024-0330/full/html

Eshbayev, O., Xursandov, K., Pulatovna, K. U., Sitora, A., & Jamalova, G. (2024). Advancing green technology systems through digital economy innovations: A study on sustainable. *E3S Web of Conferences*, *576*, 02009. https://www.e3s-conferences.org/articles/e3sconf/abs/2024/106/e3sconf_eic2024_02009/e3sconf_eic2024_02009.html

Fan, Z., Yan, Z., & Wen, S. (2023). Deep learning and artificial intelligence in sustainability: A review of SDGs, renewable energy, and environmental health. *Sustainability*, *15*(18), 13493.

Fikri, Y., & Rhalma, M. (2024). Artificial Intelligence (AI) in early childhood education (ECE): Do effects and interactions matter? *International Journal of Religion: Politics, Sociology, Culture*, *5*(11), 7536–7545.

Fikri, Y., Talaat, R., Shaheen, A., Hassan, A., & Khataan, A. (2025). Through the ARDL approach: Is there a nexus between renewable energy consumption, economic growth, and foreign direct investment in the Moroccan context? *Sustainability*, *17*(21), 9762.

Goel, A., Masurkar, S., & Pathade, G. R. (2024). An overview of digital transformation and environmental sustainability: Threats, opportunities, and solutions. *Sustainability*, *16*(24), 11079.

Ibrahim, B. S. K. K., Sutcu, M., & Al Momany, A. (2024). Advancing Sustainable Development through the Integration of AI, IoT, and Robotics. *Evolution of Information, Communication and Computing System*, 19–49.

Kataria, A., Rani, S., & Kautish, S. (2024). Artificial Intelligence of Things for Sustainable Development of Smart City Infrastructures. In W. Leal Filho, S. Kautish, T. Wall, S. Rewhorn & S. K. Paul (Eds.), *Digital Technologies to Implement the UN Sustainable Development Goals* (pp. 187–213). Springer Nature Switzerland. https://doi.org/10.1007/978-3-031-68427-2_10

Keskar, A. (2022). Harnessing IoT and AI for driving sustainability: Advanced frameworks for smart resource management and decarbonization. *International Journal of Enhanced Research in Management & Computer Applications*, *11*, 27–37.

Kumar, A. A., Suryadevara, R., Sowmyya, T., & Chanal, G. B. (2024). Exploring the Intersection of AI, Healthcare, and Environmental Sustainability: Future Trends and Challenges. In R. Singh, W. Shafik, D. Crowther & V. Kumar (Eds.), *Transforming Healthcare Sector Through Artificial Intelligence and Environmental Sustainability* (pp. 23–47). Springer Nature Singapore. https://doi.org/10.1007/978-981-97-9555-0_2

Mahmood, H. S., Abdulqader, D. M., Abdullah, R. M., Rasheed, H., Ismael, Z. N. R., & Sami, T. M. G. (2024). Conducting in-depth analysis of AI, IoT, web technology, cloud computing, and enterprise systems integration for enhancing data security and governance to promote sustainable business practices. *Journal of Information Technology and Informatics*, *3*(2), 297–332.

Mak, S.-L., Wong, Y.-M., Ho, K.-C., & Lee, C.-C. (2022). Contemporary green solutions for the logistics and transportation industry—With case illustration of a leading global 3PL based in Hong Kong. *Sustainability*, *14*(14), 8777.

Mallo, S. F., Abdulqader, D. M., Abdullah, R. M., Ismael, H. R., Rashid, Z. N., & Sami, T. M. G. (2024). A review on feasibility of web technology and cloud computing for sustainable ES: Leveraging AI, IoT, and security for green operations. *Journal of Information Technology and Informatics*, *3*(2). https://www.researchgate.net/profile/Dildar-Abdulqadir/publication/382939604_A_Review_on_Feasibility_of_Web_Technology_and_Cloud_Computing_for_Sustainable_ES_Leveraging_AI_IoT_and_Security_for_Green_Operations/links/66bdacc02ff54d6c9ed0f87f/A-Review-on-Feasibility-of-Web-Technology-and-Cloud-Computing-for-Sustainable-ES-Leveraging-AI-IoT-and-Security-for-Green-Operations.pdf

Mohsen, B. M. (2024). Ai-driven optimization of urban logistics in smart cities: Integrating autonomous vehicles and IoT for efficient delivery systems. *Sustainability*, *16*(24), 11265.

Nwaigbo, J. C., Sanusi, A. N., Akinode, A. O., & Cyriacus, C. (2025). Artificial intelligence in smart cities: Accelerating urban sustainability through intelligent systems. *Global Journal of Engineering and Technology Advances*, *24*(03), 051–073.

Ogbanga, M. M., Sharma, S. N., Pandey, A. K., & Singh, P. (2025). Artificial Intelligence in Social Work to Ensure Environmental Sustainability. In M. Nasr, A. Negm & L. Peng (Eds.), *Artificial Intelligence Applications for a Sustainable Environment* (pp. 491–508). Springer Nature Switzerland. https://doi.org/10.1007/978-3-031-91199-6_16

Omar, H. Y., Abdulqader, D. M., Abdullah, R. M., Ismael, H. R., Rashid, Z. N., & Sami, T. M. G. (n.d.). A review on upshots of cloud computing and web technology on the future green transformation: AI, IoT, and secure enterprise systems in fostering sustainable work practices. *Journal of Information Technology and Informatics*, *3*. Retrieved January 22, 2026, from https://www.researchgate.net/profile/Zryan-Najat/publication/382941747_A_Review_on_Upshots_of_Cloud_Computing_and_Web_Technology_on_the_Future_Green_Transformation_AI_IoT_and_Secure_Enterprise_Systems_in_Fostering_Sustainable_Work_Practices/links/67726fed894c55208534a8a2/A-Review-on-Upshots-of-Cloud-Computing-and-Web-Technology-on-the-Future-Green-Transformation-AI-IoT-and-Secure-Enterprise-Systems-in-Fostering-Sustainable-Work-Practices.pdf

Rawat, K. S., & Yadav, M. (2025). Decoding the intellectual landscape and evolving research paradigm of artificial intelligence in environmental sustainability. *IEEE Internet of Things Journal*. https://ieeexplore.ieee.org/abstract/document/11079945/

Sanusi, A. N., Bayeroju, O. F., & Nwokediegwu, Z. Q. S. (2023). Framework for leveraging artificial intelligence in monitoring environmental impacts of green buildings. *International Journal of Advanced Multidisciplinary Research and Studies*, *3*(1), 1194–1203.

Ullah, F., Olatunji, O., & Qayyum, S. (2024). A scoping review of green Internet of Things in construction and smart cities: Current applications, adoption strategies and future directions. *Smart and Sustainable Built Environment*. https://www.emerald.com/insight/content/doi/10.1108/sasbe-11-2023-0349/full/html

Wu, H., Liu, J., & Liang, B. (2024). AI-driven supply chain transformation in industry 5.0: Enhancing resilience and sustainability. *Journal of the Knowledge Economy*, *16*(1), 3826–3868. https://doi.org/10.1007/s13132-024-01999-6

Yigitcanlar, T., Desouza, K. C., Butler, L., & Roozkhosh, F. (2020). Contributions and risks of artificial intelligence (AI) in building smarter cities: Insights from a systematic review of the literature. *Energies*, *13*(6), 1473.

13 AI and IoT-Enabled Smart Classrooms

Redefining Teaching–Learning Processes in Higher Education

Shyamasundar Tripathy, Sorie Saidu Kamara, B. Vamsi Krishna, V. S. Prasad Kandi, and Umesh Kumar Shukla

13.1 INTRODUCTION

The pace of rapid development in technology is transforming the higher education domain, mainly led by digital, artificial intelligence (AI) and Internet of Things (IoT). Driven by such technology diffusion, smart environments and technologies have promoted the creation of smart classrooms in educational institutions for enhancing teaching efficiency, students' engagement and learning achievement [1]. Smart classrooms integrate AI-driven analytics, smart LMS and IoT-based devices such as smartboards, sensors, connected learning platforms to facilitate the interactive adaptive and data-informed education [2]. AI in education also supports personalized learning by means of adaptive content delivery, automated grading, predictive analytics, and just-in-time feedback to tailor instruction to students' diverse learning needs [3]. IoT acts as an adjunct to AI, as it is useful in ensuring effortless interconnections between physical and digital learning resources which allow the real-time monitoring of classroom activities, student attendance and environmental situations [4,5]. Composed, AI and IoT form an intelligent learning ecosystem for near real-time assessment, collaborative learning, and evidence-informed teaching decisions [6]. Internationally, smart classroom approaches are being implemented in higher education as a response to challenges including student diversity, the demand for flexible learning and to improve academic performance and institutional effectiveness [7,8]. But it is not all sunshine and roses when AI gets added to IoT for education [9]. Challenges with digital infrastructure, faculty preparedness, data privacy, ethical handling of student data and lack of equal access to technology continue to be critical issues faced especially by developing countries. In the context of India, various steps like the launch of Digital India, National Education Policy (NEP) 2020 and scaling up EdTech platforms have hastened the implementation of digital technologies in higher education [9,10]. Some states like Andhra Pradesh have heavily invested in digital classrooms, smart boards and on-line educational platforms throughout the public and private sectors. However, evidence on how AI- and IoT-based smart classrooms are actually affecting the teaching–learning process at the institutional level is scant, especially in terms of primary data [11,12]. This lacuna is filled by the present study through an analysis of AI and IoT-facilitated smart classrooms in re-shaping teaching-learning processes in Higher Education Institutions (HEIs) of Andhra Pradesh, India. Through an examination of faculty and student experiences, the research contributes a micro-level perspective on patterns of adoption, perceived benefits, and implementation challenges. This kind of evidence is essential to enable the decision-making processes in institutions,

 DOI: 10.1201/9781003767855-13

design policies and strategies, and sustainable implementations of smart-classroom technologies within higher education systems.

13.2 LITERATURE REVIEW

The fusion of AI and IoT has become one of the characteristics of digital transformation in higher education [13]. Together, these technologies support the development of smart classrooms through integration of intelligent data analytics with connected physical and virtual learning spaces. Unlike traditional classrooms, AI and IoT-based smart classrooms facilitate personalized learning, coaching feedback, data-driven pedagogical decisions to transform the teaching–learning system at both the classroom and institutional levels.

13.3 CONCEPTUAL FOUNDATIONS OF AI AND IoT IN EDUCATION

When we think about AI in the context of Education, primarily it is focused on application of machine learning algorithms, NLP and data analytics to model learner behavior, personalize education, and create process efficiencies within academic environments [14]. Previous research indicates that AI-empowered learning systems serve as cognitive support tools of the instructor by diagnosing learners in need and then tailoring instructional strategies [15] From a constructivist standpoint, AI-assisted tools make possible learning through engagement by personalizing the content for learner and delivering them with timely feedback.

In contrast, the IoT emphasizes embedding sensors, devices and networked technologies in physical learning environments [16]. Classrooms that are IoT-enabled will take measurements of real-time data covering environmental, attendance and engagement levels, classroom interactions, etc. [17]. When connected with AI analytics, IoT architecture turns static classrooms into adaptive learning environments that can accommodate both F2F and blended learning approaches [18].

13.4 AI-DRIVEN TEACHING–LEARNING PROCESSES

A vast literature has discussed the value of AI helping to improve core teaching–learning activities. Adaptive learning and similar intelligent tutoring systems will adapt the metaphor of individualized pathways to fit student performance, addressing individual differences in learning. Auto-grading tools offer immediate feedback on completed tasks freeing up instructors to concentrate more on higher-level pedagogical activities [19].

Predictive learning analytics also help institutions to identify at-risk students and intervene early, which leads to better academic retention and performance [20]. Nevertheless, academics remind us to think of AI as a pedagogical tool rather than a human substitute because good teaching still depends on contextual judgment, empathy, and academic guidance [21].

13.5 ROLE OF IoT IN SMART CLASSROOM ENGAGEMENT

IoT technology in classroom allows interactivity and engagement by connecting students and teachers on the Internet to suit learning resources. Smartboards may integrate with other devices and classroom sensors to support collaborative learning activities and generate feedback to teachers. Studies show that an environment like this encourages engagement and promotes peer to peer education, which is significant for student centered learning [22]. Moreover, IoT systems can be used in education for classroom management through attendance marking and tracking learning activities for better effectiveness of instruction [23]. However, readings note that if these systems are not well supported, maintained, or updated "the IoT infrastructure can be used less or be too disruptive to the teaching function."

13.6 PEDAGOGICAL OUTCOMES AND LEARNING EFFECTIVENESS

Numerous empirical findings indicate positive associations between smart classroom acceptance and pedagogical results. It is widely reported that student engagement and collaborative learning, as well as conceptual understanding are influenced positively [24,25]. AI and IoT tools are considered useful aids in enhancing effectiveness of teaching and instructional planning by the faculty.

However, literature also indicates that the pedagogical benefits are contingent upon faculty digital skills and instructional design behaviors. Technology application that is not pedagogically aligned is – functional adoption where learning gains are limited [26].

13.7 IMPLEMENTATION BARRIERS AND ETHICAL DIMENSIONS

While AI and IoT-based classrooms appear promising, they have to overcome certain challenges before being implemented. The lack of the proper infrastructure, high cost and not trained faculty are cited as common challenges [27]. There are also ethical issues around data privacy and student surveillance, algorithmic bias, as well as transparency which have become particularly topical [28]. These are aspects that make the need for inbuilt institutional policies to balance innovation with appropriate ethical governance.

13.8 INDIAN HIGHER EDUCATION CONTEXT AND RESEARCH GAP

AI and IoT adoption in higher education are highly influenced by the national policy initiatives, that are NEP 2020 and Digital India for Indian context. Current literature in this realm focuses on strategical paradigm and potential more than empirical evaluation of classroom-level outcomes [29]. Few empirical studies based on such primary data are available, especially at state level.

Further, the majority of Indian research tends to cover the student perspective only or that of institutional preparedness, and rarely is there an attempt to integrate both student and faculty perspectives in a single analytical model [30]. Also, the infrastructural and digital preparedness at regional level also requires localized research. There is also a paucity of rigorous empirical insights from states such as Andhra Pradesh regarding the operation of AI-IoT-fused smart classrooms in actual university settings.

Filling these gaps, this research offers a holistic applied and empirical analysis of AI and IoT integrated smart classrooms in higher educational institutes in the context of Andhra Pradesh. Through combining technological, pedagogical, and user perspectives, the study not only provides descriptive views but also prescriptive insights that may enrich theory building and inform best practice.

13.8.1 Objectives of the Study

1. To study the usage of AI and IoT based smart classroom technologies in HEIs of Andhra Pradesh.
2. To investigate the influence of AI and IoT-based smart classrooms on teaching efficiency and student participation.
3. To understand faculty and student expectations on smart classroom tools.
4. To analyze the issues and barriers in the adoption of AI and IoT based smart classrooms on higher education sector.

13.8.2 Hypotheses

H1: There is a positive and significant impact of the adoption of smart classroom technologies enabled by AI and IoT on teaching effectiveness in HEIs.

H2: Smart classrooms using AI and IoT have a significant positive impact on students engagement and participation in the learning process.

H3: People (faculty and students) have positive attitudes toward integrating AI and IoT technology into a smart classroom.

H4: The constraints of infrastructure and training have a significant influence on the efficient operation of AI and IoT-based smart classrooms in HEI's.

13.9 RESEARCH FRAMEWORK

Figure 13.1 presents the theoretical framework that explains how AI and IoT-based smart classroom adoption affects teaching–learning outcomes in HEIs. Smart institutional adoption: The model situates smart classroom diffusion at its focal point as the incorporation of intelligent devices and networked technologies in classroom functions.

User perception (faculty and students) is identified as a major enabler. Positive attitudes, perceived utility and acceptance to AIoT enhance the successful adoption and meaningful classroom use. The direct impact of smart classroom use is found in teaching effective improvement and student engagement increase, especially regarding enriched lecture style, improved interaction and more active learning.

These two dimensions together add to the overall teaching–learning achievement, which in turn is manifested as improved learning quality and better academic performance. In the meanwhile, challenges of implementation including lack of adequate infrastructure, training and privacy issues are represented as constraining factors that can inhibit these effects.

In conclusion, the model suggests that the effect of smart classroom success is not only transmitted through technology adoption but also reinforced by user readiness and controlled by the degree at which institutions can effectively handle implementation issues.

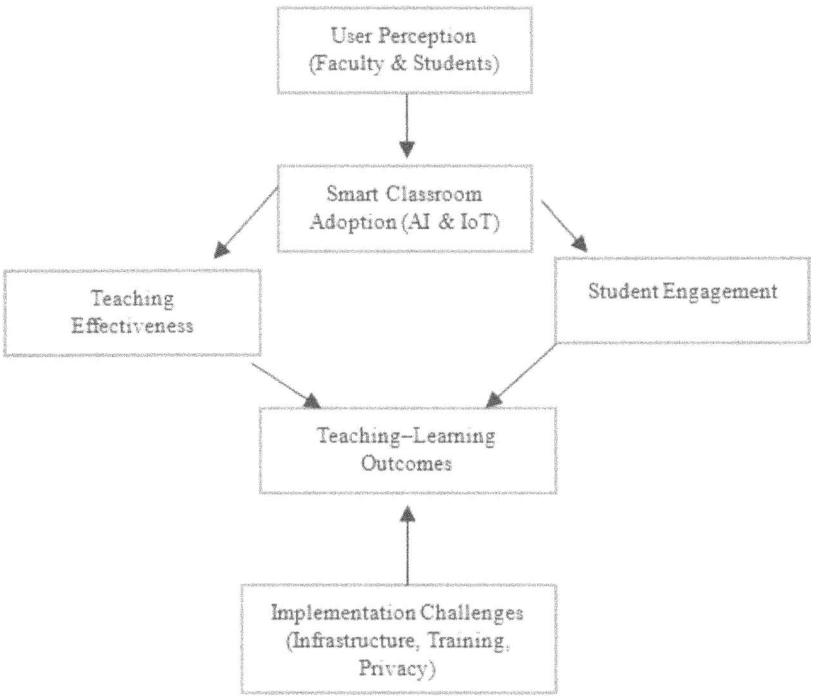

FIGURE 13.1 Smart classroom adoption and teaching–learning outcomes framework.

13.10 RESEARCH METHODOLOGY

The study is conducted with descriptive and important research design to understand the role of AI and IoT-enabled smart class rooms on teaching learning processes in higher educational institutions in Andhra Pradesh, India. The research is primary data–driven and attempts to provide institution-level evidences on adoption, effectiveness, challenges of smart classroom technologies.

The study population includes the faculties and students of public, and private universities (government/government-aided) and affiliated colleges of Andhra Pradesh which introduced AI and IoT based smart classroom facilities. To guarantee equitable representation of the key stakeholders in the teaching–learning process, a stratified random sampling approach has been used. Faculty and students defined two strata. This was justified as while perceptions about technology use and its effect on learning may vary significantly between teachers and students, stratifying the sample allows for an equal representation of both groups.

This achieved sample consisted of 300 participants, which was judged to be a sufficient number, according to earlier empirical educational technology research and the needs for regression-based analysis. The data were obtained from 120 faculty members and 180 students, which provided adequate variability, as well as statistical power to test the study hypotheses. The sample size was also reasonable considering feasibility considerations, as well as access to institutional resources available at the time and the requirement for minimal loss to follow up in a multicenter state-wide cohort.

Data for the present study were collected by a structured questionnaire that had made to obtain primary data. The questionnaire comprised items in demographic background, AI and IoT base smart classroom application coverage, teaching effectiveness, student's engagement, user perceptions and challenges of implementation. The measurement items were taken from established scales in available literature and modified to suit the college environment of Andhra Pradesh. Responses were provided on a Likert scale of "strongly disagree" to "strongly agree."

A pilot testing of questionnaire was performed on a small sample of study subjects in order to establish readability, dependability and content validity. Slight changes were made based on pilot input. The internal consistency estimates (Cronbach's alpha) demonstrated the reliability of the constructs which were deemed adequate. Analyses were conducted using statistical software. Simple descriptive statistics were used to characterize characteristics of respondents and level their technology adoption while inferential statistical tools such as Correlation, Multiple Regression were employed to assess the hypothesized relationships. The ethical issues of informed consent, anonymity and maintaining the confidentiality of respondents' answers were observed during the study.

13.11 RESULTS AND DISCUSSION

This section describes the empirical results of the study, and analyses them in terms of the research model and hypotheses (H1–H4). The study was made of a primary nature, which drew its data from Faculties and Students in the HEI of Andhra Pradesh. The synthesis consists of descriptive and inferential findings to describe the way AI-based IoT guided smart classrooms impact teaching-the learning processes.

13.12 RESPONDENT PROFILE

To interpret the results, it is important to understand the characteristics of the sample since perception toward smart classroom technology may diffuse across roles and institutional context. Faculty and students are two separate stakeholders in the teaching–learning space, thereby inclusion of both types strikes a balance for analytic generalizability. Second, the sample includes respondents from both public and private institutions which enables representation of institutional diversity in infrastructure resources, administrative support, and technology integration experience.

TABLE 13.1
Profile of Respondents (n = 300)

Category	Frequency	Percentage
Faculty	120	40
Students	180	60
Public Institutions	138	46
Private Institutions	162	54

As Table 13.1 demonstrates, the sample is well distributed among faculty and students, which provides the degree to which both teaching and learner perspectives are adequately represented. Fairly proportional representation of public and private institutions increases the applicability of findings across various governance and funding models in Andhra Pradesh's higher education system.

13.12.1 Adoption of AI and IoT-Enabled Smart Classrooms

To gauge the penetration level of smart classroom, one would need to understand how much AI enabled tool and IoT-based infrastructure are being used in educational process. Adoption rates reveal organizational preparedness; they also indicate to what degree innovative technologies have transcended the experimental phase within academe. A higher average score indicates more comprehensive integration into teaching–learning activities, and the variability demonstrates the unevenness of access and adoption by organizations.

Table 13.2 reveals that the use of AI-supported educational tools is somewhat higher than IoT infrastructure. This could mean that applications requiring minimal hardware, such as learning analytics-driven systems, online testing and adaptive content session rather than more traditional heavy-hardware IoT-based solutions are easier to adopt. The diversity of responses suggests discrepancies in resources, technical support and infrastructure among institutions in Andhra Pradesh.

13.12.2 Teaching–Learning Outcome Measures

To better understand the effects caused by the adoption of smart classrooms, it is called for to study its pedagogical outcomes. Teaching quality and pupil engagement were identified as primary outcome variables since these directly measure gains in instructional quality and student participation. Larger average values of these dimensions suggest that smart classes are not just technological accessories but they have a significant contribution to the academic activities.

Table 13.3 shows high mean values for both outcome variables, indicating a high level of perceived positive influence on AI and IoT empowered smart classrooms. In the view of teachers, smart classrooms enhanced the presentation of lessons, feedback and assessment. Engagement levels were also reported as being higher in terms of interactive content, teamwork, and active participation that suggested a move toward a more active learning environment.

TABLE 13.2
Adoption Levels of Smart Classroom Technologies

Variable	Mean	Std. Deviation
AI-enabled educational tools	3.92	0.71
IoT-enabled classroom infrastructure	3.85	0.74

TABLE 13.3
Teaching–Learning Outcome Measures

Outcome Variable	Mean	Std. Deviation
Teaching effectiveness	4.01	0.68
Student engagement	3.96	0.72

Source: Authors.

13.12.3 Hypothesis Testing Results

To test the proposed relationships in the research framework empirically, multiple regressions were performed. This research reveals the direct impact of smart classroom adoption on teaching–learning, indirect effect through user perception and finally adoption challenges. The regression results empirically verify or deny the proposed hypotheses.

Table 13.4 demonstrates the findings from hypothesis testing on the association between smart classroom adoption, user perception, implementation obstacles and teaching–learning outcomes. The values of the regression coefficients, and hence their levels of statistical significance, offer crucial information relative strength and direction of such relationships within higher academe.

The findings of H1 show that smart classroom adoption is significantly positive to teaching effectiveness ($\beta = 0.47$, $p < 0.01$). This suggests that the convergence of AI and IoT technologies has effects on instruction delivery, assessment practices and classroom arrangement. With real-time information, we make better instruction personalization possible and more accurate monitoring what students are doing with the information that creates for us more responsive learning and hence better teaching.

Likewise, H2 has a very strong and positive connection between the acceptance of smart classroom and student engagement ($\beta = 0.43$, $p < 0.01$). This result suggests that interactive technology, real-time feedback systems and connected learning environments promote active involvement, work-based learning activities and learner engagement. The relatively high beta value indicates that smart classroom tools play an important role in creating a learning environment full of engagement.

The findings for H3 suggest that human perception is a vital factor with respect to the utilization of smart classroom technology ($\beta = 0.39$, $p < 0.01$). The degree to which faculty and students have positive attitudes toward AI and IoT technologies does influence their level of adoption and meaningful integration of such tools in teaching/learning paradigms. This highlights the fact that only infrastructure does not guarantee technological effectiveness, and users' acceptance, confidence as well as perceived usefulness matters the most.

However, H4 has a significant inverse association between challenges of implementation and teaching–learning outcomes ($\beta = -0.28$, $p < 0.05$). This suggests that obstacles such as poor infrastructure, weak technical support, inadequate training and concerns about data privacy impede the

TABLE 13.4
Results of Hypothesis Testing

Hypothesis	Relationship Tested	β	p-value	Result
H1	Smart Classroom Adoption → Teaching Effectiveness	0.47	<0.01	Supported
H2	Smart Classroom Adoption → Student Engagement	0.43	<0.01	Supported
H3	User Perception → Effective Utilization	0.39	<0.01	Supported
H4	Implementation Challenges → Teaching–Learning Outcomes	−0.28	<0.05	Supported

positive influence of smart classroom adoption. While these are generally moderate in effect size, it highlights the need to mitigate institutional and operational barriers if the benefits of AI and IoT enabled learning environments are to be maintained.

In general, as seen in Table 13.4, the findings highlight that the increase of smart classroom adoption enhances teaching performance and students' participation, though these advantages are subject to good user perception and overcoming implementation difficulties.

13.13 DISCUSSION

The findings of the study provide strong empirical evidence that AI and IoT add on smart classrooms are significantly influencing the teaching–learning activities in higher educational institutions of Andhra Pradesh. The positive and significant relationship found between smart classroom usage, teacher performance and student engagement is the evidence of morphing nature of intelligent and connected learning environments. The results contribute to the nascent literature for promoting AI and IOT as enablers for data-driven, learning-centered pedagogy at higher education.

Notably, the large impact of smart classroom usage on teaching quality suggests AI-facilitated tools such as automated testing, learning analytics and intelligent tutoring systems augment faculty ability to design, deliver and evaluate instruction in a better way. At the heart of this process are IoT-enabled campuses, which enable real-time interaction, improved learning outcomes and blurring boundaries between the digital and physical classrooms. In resource-constrained environments such as in Andhra Pradesh, where teacher-student ratios are high and resources limited, it seems that these tools may be useful for enhancing teaching quality and monitoring academic achievement. Similarly, high level of student engagement means that smart classrooms will stimulate the efficient cooperating and the long-term concentration of students. Interactive screens, devices that read/y linked to each other and AI-assisted feedback loops that kick in so that the students don't just passively listen, but participate through experiential learning. This is aligned with contemporary pedagogical approaches, which suggest that engaging learners in an activity supports enhanced learning while working within technology-supported environments.

The strong influence of faculty and student perceptions as a cause is an evidence for the human factor as one of the dimensions in adopting educational technology. Perceived usefulness, perceived ease of use and relevance, both support real using purposes for AI tools and IoT devices. This supports technology acceptance and diffusion theories, showing that institutional investments in infrastructure alone are insufficient for the sustainable cycle of change, with the secondary building of capacity, ongoing training and changing management still needing to be supported. Even when the new technologies arise from the classroom, if they are not pedagogically consistent, and users do not trust them there is a risk of smart classrooms never reaching full potential. At the same time, barriers to adoption are impacting the efficacy of smart classrooms broadly. Negative aspects of AI and IOT adoption in the form of lack of infrastructure, faculty support, internet limitations, development programs for faculties and data privacy/surveillance were recognized. These are manifestations of the structural and systemic malaise within higher education, and are rife in developing countries in particular. In Andhra Pradesh, inequity between urban and semi-urban schools deepens these constraints.

Overall, the discussion emphasizes that AI and IoT-based smart classrooms possess tremendous pedagogical potential, but this depends on institutional readiness, governance principles and ethical considerations. The successful integration of technology must subsequently be supported by a well-rounded components of innovation, policy-making, teaching force development and inclusive strategies for implementation. Findings from these study, thus contribute to enrich the global digital transformation on higher education discourses with an invaluable domestic knowledge.

13.14 CONCLUSION

The study sheds light on the impact of AI and IoT-tool assisted-smart-class rooms at higher educational institutions (of AP, India) on teaching–learning process with first hand data collected through a survey from faculty members and students. The observed empirical evidence contributes to and offers new evidence on the relation between the implementation of smart classroom technologies, university teaching effectiveness, and student engagement. Conversely, AI-driven tools for data-informed instruction planning and assessments and personalized feedback get their mojo with class interactivity, connection and the like that make up learning space is supercharged with IoT-enabled infrastructure.

The findings also highlight the significance of human and organizational aspects in influencing successful roll out of smart classroom initiatives. A further driver for the successful use of AI and IoT technologies were reported to be positive faculty [and] student attitudes – recommendation: Investments in technology need to go hand-in-hand with appropriate training-supported-users. Comparatively, implementation issues like infrastructure limitation, technical support, privacy concerns on data followed by adoption of smart classroom lead to a decrease in its benefits. This research also contributes to business and social value debate by providing empirical evidence at state level in a context where much work has been limited to conceptual debates and national policy frameworks. The results highlight the need for a holistic approach to digital transformation among managers of HEIs. Overall, the implications of the findings from this research are that smart classrooms have a capacity to revolutionize teaching–learning in higher education when used strategically and aligned with broad-spectrum educational goals.

13.15 LIMITATIONS AND FUTURE RESEARCH DIRECTIONS

Notwithstanding its empirical value, this study has some limitations that need to be considered in interpreting the results. First, the study is confined to higher educational institutions of Andhra Pradesh with geographical restrictedness. While such an orientation at the state level offers useful context, however, findings of the study may not be readily generalizable to other parts of India or to other nations where the institutional preparedness for e-governance and policy structures as well as technology infrastructure might differ. Second, the study is based on self-reported data from a questionnaire, which may be influenced by reporting bias, social desirability effects and subjective measures of technology use and effectiveness. Third, the cross-sectional nature of our research design only provides a snapshot view of teachers' perceptions and does not consider how technological adoption, or pedagogical outcomes, may change over time. Finally, although the paper addressed some of them (e.g., technological, perceptual, and institutional issues) it did not explicitly report on other key factors (e.g., organizational culture or leadership support and long-term learning results) that are essential to make things work. Discussion Schools with a Learning Management System trained their teachers more successfully than schools without.

There are several areas that novelty and future research can address these limitations. Long-term/follow-up studies of the effects of AI and IoT-based smart classroom can be conducted after a lapse to evaluate how the Smart Classroom system impact change as the institutions grow more experienced in their exploiting technology. Cross-sectional studies in various Indian states or simultaneous comparison between public and private institutions would provide further understanding on regional and institutional variations. Mixed methods and other research in the future may embed these quantitative scales with qualitative interview or classroom observation, to complement scores and collect sound evidence of subtler pedagogical practices. Furthermore, it is possible to test more complicated relationships between technological-pedagogical-organizational (TPO) factors using advanced statistical methods including path analysis or structural equation modeling. Further analysis of the ethical governance, data privacy frameworks and long-term academic impacts through AI and IoT adoption are also some of the possible avenues for further research.

REFERENCES

1. Mohamed Hashim, M. A., I. Tlemsani, and R. Matthews, "Higher education strategy in digital transformation," *Education and Information Technologies*, vol. 27, pp. 3171–3195, Apr. 2022, doi: 10.1007/S10639-021-10739-1
2. Kurni, M., and K. G. Srinivasa, Technology-based learning experiences with IoT in education. In *The Internet of Educational Things*, Springer, pp. 113–133, 2025, doi: 10.1007/978-3-031-67387-0_8
3. Sundari, M. S., H. R. Penthala, and A. Nayyar, Transforming education through AI-enhanced content creation and personalized learning experiences. In *Impact of Artificial Intelligence on Society*, pp. 98–118, Chapman and Hall/CRC, Sep. 2024, doi: 10.1201/9781032644509
4. Kumar, A., R. Kumar, A. Kumar, P. D. Yelpale, and A. Thakur, Next-Generation online education integrating AI and IoT for superior management and evaluation. In Ramanujam, E. and Chakraborty, C. *Artificial Intelligence and IoT in Online Education Systems*, Wiley, pp. 27–62, Dec. 2025, doi: 10.1002/9781394302666.CH2
5. Kurni, M., and K. G. Srinivasa, IoT for Education, In *The Internet of Educational Things*, Springer, pp. 1–22, 2025, doi: 10.1007/978-3-031-67387-0_1
6. Wangoo, D., and S. R. N. Reddy, "Smart learning environments framework for educational applications in IoT enabled educational ecosystems: A review on AI based GUI tools for IoT wearables," *IEEE 17th India Council, and Undefined*, Accessed: December 23, 2025, https://ieeexplore.ieee.org/abstract/document/9342150/
7. Acharya, S., "Beyond learning outcomes: impact of organizational flexibility on strategic performance measures of commercial e-learning providers," *Global Journal of Flexible Systems Management*, vol. 20, no. 1, pp. 31–41, Mar. 2019, doi: 10.1007/S40171-018-0199-3
8. Müller, C., T. Mildenberger, and D. Steingruber, "Learning effectiveness of a flexible learning study programme in a blended learning design: Why are some courses more effective than others?" *International Journal of Educational Technology in Higher Education*, vol. 20, no. 1, Dec. 2023, doi: 10.1186/S41239-022-00379-X
9. Ahuja, K., and I. Bala, Role of artificial intelligence and IoT in next generation education system. In Al-Turjman, F., Nayyar, A., Devi, A., Shukla, P. K. (eds.), *Intelligence of Things: AI-IoT Based Critical-Applications, and Undefined*. Springer, pp. 189–208, Oct. 2021, doi: 10.1007/978-3-030-82800-4_8
10. Nagaraju, S., and J. V. Gorabal, Advances in Teaching–Learning technologies: transforming education in the digital age. In *Work Integrated Learning-Directions for the Future*, Springer, vol. 1206, pp. 601–620, 2025, doi: 10.1007/978-981-96-0201-8_38
11. Marfo, B., F. K. Bondinuba, and C. M. Mewomo, "A theoretical framework towards leveraging Internet of Things applications in building energy efficiency," *Property Management*, pp. 1–24, Nov. 2025, https://www.emerald.com/pm/article-abstract/44/2/232/1318001/A-theoretical-framework-towards-leveraging?redirectedFrom=fulltext
12. Huang, Z., "IoT-inspired teaching for legal education: AI-based learning based on decision tree algorithm." *Soft Computing*, vol. 28, no. 2, pp. 1609–1631, Jan. 2024, doi: 10.1007/S00500-023-09451-8
13. Koong L., H.-C. Yang Lu, V. K. Quy, B. T. Thanh, A. Chehri, D. M. Linh, and D. A. Tuan., "AI and digital transformation in higher education: Vision and approach of a specific university in Vietnam," *Sustainability*, vol. 15, no. 14, p. 11093, 2023, doi: 10.3390/su151411093
14. Sajja, R., Y. Sermet, D. Cwiertny, and I. Demir, "Integrating AI and learning analytics for data-driven pedagogical decisions and personalized interventions in education," *Technology, Knowledge and Learning*, 2025, doi: 10.1007/S10758-025-09897-9
15. Chakraborty, S., B. Misra, and N. Dey, *AI-Empowered Knowledge Management in Education*. 2024. Accessed December 24, 2025, https://link.springer.com/content/pdf/10.1007/978-981-97-2574-8.pdf
16. Marquez, J., J. Villanueva, Z. Solarte, and A. Garcia, "IoT in education: Integration of objects with virtual academic communities," *Computer Science, Education, Engineering*, vol. 444, pp. 201–212, 2016, doi: 10.1007/978-3-319-31232-3_19
17. Haque, S., M. Alimul Haque, D. Kumar, K. Mishra, F. Islam, S. Ahmad, K. Kumar, and B. Kumar Mishra, "Assessing the impact of IoT enabled E-learning system for higher education," *SN Computer Science*, vol. 4, no. 5, 2023, doi: 10.1007/S42979-023-01860-8
18. Lomer, S., and E. Palmer, "'I didn't know this was actually stuff that could help us, with actually Learning': Student perceptions of Active Blended Learning," *Teaching in Higher Education*, vol. 28, no. 4, pp. 679–698, 2023, doi: 10.1080/13562517.2020.1852202
19. Tamaki, K., and K. Ishii, Automation in education, training, and learning systems. In *Springer Handbook of Automation*, vol. Part F674, pp. 1349–1370, 2023, doi: 10.1007/978-3-030-96729-1_63

20. Foster, E., and R. Siddle, "The effectiveness of learning analytics for identifying at-risk students in higher education," *Assessment & Evaluation in Higher Education*, vol. 45, no. 6, pp. 842–854, Aug. 2020, doi: 10.1080/02602938.2019.1682118
21. Herath, D. B., E. Ode, and G. B. Herath, "Can AI replace humans? Comparing the capabilities of AI tools and human performance in a business management education scenario," *British Educational Research Journal*, vol. 51, no. 3, pp. 1073–1096, June 2025, doi: 10.1002/BERJ.4111
22. Lewin, C., S. Cranmer, and S. McNicol, "Developing digital pedagogy through learning design: An activity theory perspective," *British Journal of Educational Technology*, vol. 49, no. 6, pp. 1131–1144, Nov. 2018, doi: 10.1111/BJET.12705
23. Tan, P., H. Wu, P. Li, and H. Xu, "Teaching management system with applications of RFID and IoT technology," *Education Sciences*, 2018, doi: 10.3390/educsci8010026
24. Mustafa, M., V. Rao Naidu, A. Mohammed, K. A. Jesrani, R. Hasan, and G. Al Hadrami, "Group engagement in cognitive activity and metacognitive regulation during collaborative learning: Can it explain differences in students' conceptual understanding?," *Metacognition and Learning*, vol. 9, no. 3, pp. 287–307, Dec. 2014, doi: 10.1007/S11409-014-9117-Z
25. Qureshi, M. A., A. Khaskheli, J. A. Qureshi, S. A. Raza, and S. Q. Yousufi, "Factors affecting students' learning performance through collaborative learning and engagement," *Interactive Learning Environments*, vol. 31, no. 4, pp. 2371–2391, 2023, doi: 10.1080/10494820.2021.1884886
26. Tarling, I., and D. Ng'ambi, "Teachers pedagogical change framework: A diagnostic tool for changing teachers' uses of emerging technologies," *British Journal of Educational Technology*, vol. 47, no. 3, pp. 554–572, May 2016, doi: 10.1111/BJET.12454
27. Gautam, A., and A. Kotiyal, Connecting the future: Cloud-based IoT in education. In *Integration of Cloud Computing and IoT*, Chapman and Hall/CRC, pp. 274–308, Jan. 2024, doi: 10.1201/9781032656694-15
28. Regan, P., and J. Jesse-Ethics, "Ethical challenges of edtech, big data and personalized learning: Twenty-first century student sorting and tracking," *Ethics and Information Technology*, vol. 21, no. 3, pp. 167–179, Sep. 2019, doi: 10.1007/S10676-018-9492-2
29. Nehru, R. S. S., T. Q. Cuong, and A. R. Prakash, Higher Education: AI Applications for Blockchain-Based IoT Technology and Networks. In *AI Models for Blockchain-Based Intelligent Networks in IoT Systems*, Springer, pp. 261–283, 2023, doi: 10.1007/978-3-031-31952-5_12
30. Bordoloi, R., P. Das, and K. Das, "Perception towards online/blended learning at the time of Covid-19 pandemic: An academic analytics in the Indian context," *Asian Association of Open Universities Journal*, vol. 16, no. 1, pp. 41–60, May 2021, doi: 10.1108/AAOUJ-09-2020-0079

14 From Authority to Advisor

How AI Co-Teaching Tools Redefine Teacher Identity

Shyamasundar Tripathy, José G. Vargas-Hernández, and Islam Elgammal

14.1 INTRODUCTION

The integration of AI in education is seen as a revolutionary change to the entire field of education through teaching and learning (Roll & Wylie, 2016). While in previous technology, teachers often took on roles of "user" or "adopter" of instructional tools, modern AI systems have the potential to be full partners in pedagogical decision-making. Such arrangements mediate knowledge, authority, and relationships not only about the content taught but also how they are managed within classroom practice (Hermansen, 2017). When AI tools co-design lessons, produce real-time instructional guidance, or prompt live feedback from students, they do more than support teaching work. They redistribute instructional authority and subvert assumptions about teachers' professional roles and expertise.

The sense of self that teachers develop through practice, the framing of which as being dynamic and socially constructed is a central feature of teacher identity, has typically been conditioned by epistemic authority, pedagogical autonomy and relational primacy in the classroom (Zembylas, 2003). They have been the schefters of knowledge, power and relationship in education. But the development of AI co-teaching tools—systems that use machine learning, natural language processing and learning analytics to take part in teaching processes—challenges these underlying principles (Topali et al., 2025).

Recent research has shed light on the rapid development of AI in education. Possaghi et al. (2025) demonstrate that multimodal learning analytics dashboards are leveraging indicators not only of student performance and engagement, but also affective states to sustain real-time pedagogic decision-making. Such systems assist teachers to identify vulnerable students, personalize the interventions for them, and uncover patterns in learning that are not obvious at first glance. Seufert et al. (2025) discuss the increasing deployment of conversational AI tools as instructional copilots to aid teachers in lesson planning, content creation, and language support. Similarly, Ojetunde and Ramnarain (2025) report IoT-supported smart classrooms where environmental and physiological sensors guide customized teaching modes. Taken together, these developments suggest a move from AI as a passive tool for support to the use of AI as an active teaching agent.

However, it is not yet clear how AI co-teaching reconfigures teacher identity, as empirical research in this area has been nascent. The current research is mainly focused on adoption, usability, and learning effects, but seems to neglect underlying professional and identity-related consequences of the delegation of instructional authority to algorithmic systems (Nopas, 2025). This divide is particularly stark in Global South settings.

To fill this gap, the current study investigates how teachers construe authority, autonomy, and professionalism in their AI-empowered classrooms. Grounded in role theory and socio-technical systems theory, the research provides concrete examples of shifting teacher identities and raises questions for teacher education, institutional governance, and ethical AI design.

DOI: 10.1201/9781003767855-14

14.2 LITERATURE REVIEW

14.2.1 Teacher Identity: Conceptual Foundations

Teacher identity is not an attribute that is stable to a degree, but rather a continuing act of sense-making in which teachers try to make sense of their narratives and relationships. It is (re)shaped by life's biography, the organizational format, and everyday social practices as a dynamic nexus of agency and structure. Traditionally, teacher identity has been based on epistemic authority, that is the authority as teachers in their subject area what they teach (Raviv et al., 2003), and didactic freedom to act pedagogically on one's own judgments. These have remained for centuries the matter of professional accomplishment and social status, not to mention a justification for being.

These institutions have been commonly used to weaken the society based on modern education system. Accountability systems, common curricula, and success criteria based on data systems have squeezed the room for teacher discretion in ways that places teachers more as implementers than creators of learning. The emergence of AI co-teaching tools represents a qualitative, and not just quantitative, shift. Unlike external policy shock, AI systems instruction themselves by generating courseware and analyzing students' data as well as suggesting pedagogical actions that previously were part of the human domain (Schiff, 2020).

Accordingly, teacher identity has to be conceived as multiply distributed across human and non-human actants. When it is analytical or evaluative tasks which are performed by algorithmic systems, teachers have the task of renegotiating authority, responsibility, and professional worth. This identity work goes beyond the adoption of tools and includes reimagining what it means to teach in a world where intelligent systems continue to influence how we enact instruction.

14.2.2 AI Co-Teaching: From Tool to Collaborator

It's simply the tools vs. the co-teachers, in educational technology. It's significant to note that in practice, the technology in a teaching space—whether it be chalk and blackboard or digital projection has continually been managed by the teacher. They are scaffolds to instruction and not for pedagogical judgment or how learning should go. It is a helping, not a playing position.

The AI co-teaching systems are substantially different than this method. As Teasley (2019) indicate that current learning analytics go beyond the mere presentation of information; they are able to analyze student behavior, predict learning risks, and promote pedagogical interventions. Along the way, these systems begin to guide teaching decisions that in the past were made largely due to teacher intuition and local context.

AI systems, of course, don't have intentions and moral sensibilities; what they do instead is to shift how authority works in classrooms. Teachers "now work with technologies that are extremely effective at processing information, but they lack the human sensitivities to context and interpersonal relationships," he said. According to Memarian and Doleck (2024), those collaboration are based on keywords: human-computer interaction, teacher-in-the-loop, complementary strengths model. The underlying strength of these models is to ensure that the analysis and being efficient could be facilitated by AI, but educational judgment can reside with educators.

14.2.3 Empirical Evidence on Identity Transformation

Empirical results indicate that the influence of AI co-teaching tools on teacher identity and professional authority is ambivalent, if not contradictory. Montobbio et al. (2024) in their study of 83 empirical works on AI and language learning published from 2019 to 2024, register task reallocation as an immediate impact. As AI tools take over mindless chores such as lesson planning, grading, and data analysis, teachers' roles will become less supervisory and judgment-based. For some teachers, this shift is a liberating escape from administrative overreach and a chance to focus on what works best for the learners. For others, it is deskilling, where what are essentially pedagogical skills become dumbed down to the work of managing algorithmic outputs.

And at the same time, a number of other studies point to the ways that collaborative AI can enhance professional identity. Tsou et al. (2024) also reported that ChatGPT instructed teachers (EMI, Taiwan) felt more confident after working with the AI and were able to teach content that they had previously not because of a language barrier. Rather than undermining authority, TEA-AGENCY offered support to teachers that enabled them to extend their instructional roles and try out new approaches.

With all its advantages, there are reports of tensions related to autonomy and authority. Nazaretsky et al. (2022) illustrate how, automated feedback systems for teacher professional improvement can frustrate teachers' professional control especially when the algorithmic evaluation does not conform to their pedagogical standards. Similarly, Kwet and Prinsloo (2020), the IoT enabled "smart classrooms" introduce new types of surveillance, which in turn lead to pressures for a teacher to explain decisions where they do not follow recommendations provided by AI, but may be the better pedagogical decision.

14.2.4 Research Gaps

Despite increasing attention, there are still key lacunae in the literature on what AI co-teaching does to teachers' identity. Most studies are based on questionnaires or retrospective interviews and do not account for how identity is negotiated in on-line, face-to-face class interactions with AI systems. Empirical attention is also scarce in how the AI design characteristics (e.g., explainability, path of automation) would shape teachers' sense of agency and authority. Furthermore, current evidence is biased toward Global North settings, while teachers' experiences in the Global South where educational AI is proliferating are less researched. Finally, existing research rarely reveals a sound theoretical connection between everyday teachers' identity work and institutional and policy arrangements on the macro level. This article seeks to fill these gaps by examining in detail, with empirical accounts, how teachers identity developed or transformed into AI-enabled classrooms.

14.3 THEORETICAL FRAMEWORK

This study employs an integrated theoretical framework combining role theory and socio-technical systems theory to analyze teacher identity transformation in AI co-teaching environments.

14.3.1 Role Theory and Professional Identity

Role theory is a valuable view through which to analyze the transformation of teacher identity in AI co-teaching in terms of role expectations, responsibilities and relationships. In sociological terms, occupational roles are not defined as fixed statuses, but are performed and negotiated constantly in interaction (Chu et al., 2025). The identity as teacher is constructed gradually over time, as teachers internalize these situational expectations for roles which stabilize through continual practice and feedback in the community. Role-taking is the vehicle to help teachers internalize institutional norms, and role-making provides direction for adapting the role to personal values and teaching styles.

Co-teaching in AI brings up new types of role confusion and role ambiguity. Blurred lines between human and algorithmic responsibility also pose questions of accountability, while AI-forged guidelines may run counter to a teacher's professional judgment or their pedagogical belief (Kim & Kwon, 2025). Simultaneously, AI systems generate new expectations: that teachers will be data interpreters, technology intermediators and monitors of algorithmic outputs. Role theory also brings attention to changing power relations when AI systems claim authority based on data-driven expertise, and thus teachers need to renegotiate autonomy, responsibility, and professional status within human-AI teaching arrangements in transition.

14.3.2 Socio-Technical Systems Theory

From the sociotechnical systems perspective, educational AI is considered not as isolated tools but as elements interconnected into larger technology-rich systems where social and technical factors mutually shape each other.

> If you think of it like that, the introduction of AI into classrooms will necessarily reorganize practices of teaching, professional relations and organizational routines. Technologies are not neutral: they contain values and assumptions that shape the organization of teaching and learning.
>
> **(Sony & Naik, 2020)**

AI co-teaching systems are not deployed in a vacuum. They are constituted by a variety of interlocking elements—underlying algorithms, data infrastructures, interface design, institutional policies, teacher training practices, and professional norms (Kim, 2024). Such systems may be less affected by the individual components but rather how well they work in practice within the context of teaching. Socio-technical theory tells us that technical quality doesn't hold the trump card when it comes to impact. Even a carefully designed AI system will fail if teachers are not well supported, if school cultures don't privilege data-informed practice or when the technology runs roughshod over the rhythms and constraints of everyday classroom life (Spina, 2020). Alternatively, simple AI tools can perform very well if they address actual teacher needs and are embedded in a supportive institutional framework.

This approach also raises issues of involvement and power. When decisions involving AI are made without the input of teachers, for example, a common outcome is that the systems reflect assumptions that come into conflict with those in professional practice. Socio-technical theory therefore highlights the need for participatory development, with teachers as co-designers rather than just end-users.

14.3.3 Integrated Analytical Framework

In our integrative framework, we frame teacher identity as the result of another intersection; role expectations (resulting from institutional and professional norms) with socio-technical configurations (the particular ways AI systems are fashioned, implemented and embedded in practice). This structure formulates a number of analytical propositions which inform our empirical enquiry:

- Identity change will be most salient when AI systems shift foundational role expectations, particularly of epistemic authority, pedagogical autonomy, and relational precedence. AI tools that directly engage in knowledge transmission, instructional design or assessment will cause more identity disruption than those which perform administrative duties.
- The orientation and valence of identity change will be shaped by the socio-technical design choices like explainable AI within teacher-in-the-loop workflow vs. black-box automation. Systems that show and respect how competent teachers are will help identity change to turn out positively, whereas systems that hide or belittle teacher judgment of need will be identity threatening.
- Teachers will be involved in identity work to manage role ambiguities and conflicts by building new professional self-concepts that include the algorithmic participation yet maintain certain aspects of their teaching identity. This identity work will also be influenced by career stage, previous technology experience, institutional support and professional cultures.

- Identity conversion will be flattened across fields of practice, as parts of teacher authority/expertise are amplified and others dampened. In any domain, teachers may find they are identity enhancing in one respect while threatening in another.

14.4 METHODOLOGY

14.4.1 Research Design

This study employs a convergent mixed-method research to investigate the impact of AI co-teaching on teacher identity. Such a method is suitable to the extent that identity entails quantifiable particulars and lived experiences alike. Quantitative data detail trends and associations across teachers, while qualitative data shed light on how educators make sense of and navigate identity shifts in practice.

Hence, both datasets were gathered concurrently and treated separately before being combined in the analysis. Quantitative results describe the prevalence and predictors of identity changes, whereas qualitative findings contribute to a more nuanced understanding of how and why people shift identity. As a jointly, they provide us toward the two standpoints will complement each other to offer a more balanced view about TID change in AI-assisted teaching.

14.4.2 Context and Participants

The research was carried out from August 2024 to January 2025 in government and private secondary schools and higher education institutions in Odisha and Andhra Pradesh, India. These states were selected due to the variability of educational contexts and continued implementation of educational technologies from: statewide programs, state-private partnerships. The population was secondary (Grades 9–12) and undergraduate schools offering an AI-based learning platform (e.g., lesson planning systems, automated assessment platforms, learning analytics dashboards, IoT-driven classrooms technologies) with at least one semester of exposure.

The sample of 120 teachers, chosen from institutions that fulfilled the implementation requirements is purposive. Participants had to have used AI-supported instructional tools for core academic subjects regularly (minimum three months). Teachers who used AI solely for administrative tasks were excluded. Methods The data gathering comprised surveys delivered to all the participants, and semi-structured interviews with 30 teachers chosen in order to ensure a range of experience level, area of expertise, type of institution and intensity in the use of AI. Classroom observations were also performed with a subgroup of 15 interviewees.

14.4.3 Data Collection Instruments

This study created the Teacher Identity Perception Scale (TIPS) to measure professional identity in AI co-teaching settings. The scales were developed in accordance with standard psychometric practices including writing the items based on literature and pilot interviews, expert review, pilot testing and refining. The resulting 32-item eight factor scale includes: epistemic authority, pedagogical autonomy, relational primacy, data interpretation competence, algorithmic oversight responsibility, ethical guardianship, collaborative orientation, and technological self-efficacy. Answers were given on a five-point Likert scale from strongly disagree to strongly agree.

Use of AI tools was assessed with the AI Co-Teaching Interaction Index (ACII). This tool measured the frequency and depth of AI usage through course phases, type of AI systems used, dependence on advice from AI, and perceived control over outputs by AI.

Teachers also filled out a Retrospective Authority Rating, which included ratings of their perceived authority in seven instructional domains when they were not using AI and currently. This made it possible to compare perceived changes in professional authority related to AI co-teaching.

Semi-structured interviews were utilized to obtain teachers' stories of identity change in the context of AI co-teaching. The interviews guide focused on three key domains: teacher background and use of AI tools, perceptions of role and authority pre- and post-AI adoption, experiences with trust, collaboration, and future outlook. Interviews ranged from 45 to 75 min and were completed in English or the preference of the participant (Hindi, Telugu, or Odia). All interviews were audio-taped with permission, transcribed verbatim and translated, as appropriate, into English language using back-translation for validation of data.

Fifteen teachers were observed in the classroom by means of a structured protocol. Observers attended to how and when AI systems were consulted during lessons, teacher responses to AI suggestions, mentions of or references to AI in classroom discourse and visible affective responses to AIs generated output. All observations were conducted for an entire class period and recorded in extensive field notes.

Data logs for IoT in institutions with smart classroom facilities were also harvested. These logs documented mode of use of the dashboard, type of analytics accessed, time spent interpreting AI outputs and subsequent instructional actions.

14.4.4 Data Analysis Procedures

Participant characteristics and patterns of use of the AI tool were summarized using descriptive statistics. A principal axis factoring with oblique (promax) rotation on the TIPS was conducted in order to determine dimensions of teacher identity. As EFA was used to contribute identity dimensions to the data, and due a priori anticipated relationships between identity constructs oblique rotation was used. Factor retention relied on eigenvalues > 1, scree plot examination and conceptual coherence. This was followed by removing items with low loadings or large cross-loadings. The internal consistency of the derived factors was checked by Cronbach's alpha.

Predictors of identity transformation were further analyzed by multiple regression analyses. Factor scores were treated as dependent variables, and AI usage intensity, teaching experience, type of institution, field of study and previous technology training were applied as independent variables. All the major assumptions of statistics were confirmed.

Teacher perceptions of authority in seven areas, across two time points (retrospective: pre-AI; current) were tested for differences using paired-samples t-tests. Cohen's d was used to determine effect size of change. All analyses were performed using SPSS version 28.0, and the significance level was set at $p < .05$.

Thematic analysis was used to analyze interview transcripts and observations, based on Braun and Clarke's six phase approach. The data were read and re-read, systematically coded and categorized into themes which were examined, named and redefined. A shared codebook was developed among 2 researchers after independently coding 20% of the interviews for reliability (Cohen's kappa = 0.82). Completion data were then coded in full using NVivo 14 with frequent discussions across the full dataset to ensure consistency. Levels of analysis included the surface structure and deeper structures with themes derived inductively based on theoretical perspective from the study.

The quantitative data were combined with the qualitative results at the interpretation phase, whereby identity dimensions were contrasted with emergent themes; prose was used to elucidate observed patterns and prevalence rates were used to provide context for these experiences. Ethical considerations comprised a form of informed consent, voluntary participation and the use of pseudonyms as well as anonymize data, secure storage and restricted access to research documentation.

Quantitative and qualitative findings were synthesized at the interpretation level by juxtaposing dimensions of identity with emergent themes, using narratives to understand patterns observed in response options, and using prevalence data to situate experiences. Ethical precautions were informed consent, voluntary participation, use of pseudonyms where telling was sensitive, anonymized data storage in a secure environment and limited access to research files.

14.5 RESULTS

14.5.1 Demographic Profile of Participants

Table 14.1 shows the profile of 120 teachers, who are our participants in this section, which reflects a diverse sample on some dimensions for the purpose of understanding adoption and identity transformation related to AI co-teaching.

The sample was fairly gender mixed, though women were somewhat overrepresented (56.7%), as well indicating a general feminization of the teaching labor force in India. The majority of participants fell into the 31–40 age group (43.3%); these teachers would be considered as mid-career, and early career teachers in the age group of 25–30 years were first with respect to the normal distribution. A low percentage were senior teachers, aged 51 years and older. Experience with teaching varied by career stage, the highest percentage (34.2%) of experience being between 6 and 10 years indicating that many had formed solid professional identities before AI co-teaching was introduced. Academic credentials were strong, with close to three-quarters having master's degrees and about one in five holding MPhil or PhD qualifications.

TABLE 14.1
Demographic Profile of Teacher Participants (N = 120)

Characteristic	Category	*n*	%
Gender	Female	68	56.7
	Male	52	43.3
Age range	25–30 years	28	23.3
	31–40 years	52	43.3
	41–50 years	31	25.8
	51+ years	9	7.5
Teaching experience	0–5 years	32	26.7
	6–10 years	41	34.2
	11–15 years	28	23.3
	16+ years	19	15.8
Educational qualification	Bachelor's degree	8	6.7
	Master's degree	89	74.2
	MPhil/PhD	23	19.2
Institutional type	Government secondary	35	29.2
	Private secondary	28	23.3
	Government higher Ed	31	25.8
	Private higher Ed	26	21.7
Subject area	Mathematics	24	20
	Science (physics/chemistry/biology)	31	25.8
	Languages (English/Hindi/regional)	28	23.3
	Social studies/commerce	37	30.8
Prior technology training	None	18	15
	Basic (1–2 workshops)	47	39.2
	Moderate (3–5 workshops)	38	31.7
	Extensive (>5 workshops or certification)	17	14.2
Years using AI tools	<6 months	12	10
	6–12 months	38	31.7
	1–2 years	51	42.5
	>2 years	19	15.8

The institutions are equally distributed between government and private sectors as well as between secondary and tertiary education for meaningful cross-contextual comparisons. The disciplinary diversity was provided by the different backgrounds of the subjects, with STEM, language, and social-science covered fields. Pre-existing experience of technology training was extremely diverse, with few having substantial preparation—and many minimal or no formal education at all prior to AI adoption. The majority of respondents had been using AI co-teaching tools for one or two years, indicating early but significant adoption. Overall, the sample represents teachers with established professional identities, adjusting to AI integration under non-uniform exposure and training conditions that makes a compelling case for identity shift being investigated in the following sections.

14.5.2 AI Co-Teaching Tool Usage Patterns

Teachers spent 10.4 (SD = 4.2) hours per week using AI co-teaching tools overall on average, which makes up a significant amount of their professional time and suggests that AI has become deeply embedded in teachers' day-to-day teaching practices. The intensity of use for particular instructional activities is summarized in Table 14.2.

As presented in Table 14.2, the highest amount of AI usage occurred during lesson planning and content creation with staff spending an average of 3.2 hours per week (SD = 1.8). The proportion of regular use was almost 68.3%, and high reliance was reported by 42.5%, showing that AI has been made habitual during preparatory work. They were closely followed by assessment design and grading, with 2.8 hours per week (M = 1.6) and the highest level of high reliance (51.7%), such that educationalists are becoming increasingly reliant on AI to support evaluative decisions-making.

The use of artificial intelligence for tracking students and analytics was moderate at 2.1 hours per week (SD = 1.4), with organized regularly by the activity approximately half of the time, highly organized over one third of the time. In contrast, in-class facilitation remained mostly teacher-centered (average = 1.4 h/week; SD = 1.1), where 31.7% of teachers declared regular use and only

TABLE 14.2
Usage Intensity of AI Co-Teaching Tools across Instructional Activities (N = 120)

Instructional Activity	Mean Weekly Hours (SD)	% Using Regularly	% High Reliance
Lesson planning and content generation	3.2 (1.8)	68.3	42.5
In-class instruction	1.4 (1.1)	31.7	18.3
Assessment design and grading	2.8 (1.6)	61.7	51.7
Student monitoring and analytics	2.1 (1.4)	50	35.8
Professional development	0.9 (0.8)	20	15
Overall AI usage intensity	10.4 (4.2)	–	–

AI Tool Type	n	%
Learning analytics dashboard	87	72.5
Generative AI (e.g., ChatGPT)	76	63.3
Automated assessment systems	68	56.7
AI lesson planners	45	37.5
Intelligent tutoring systems	34	28.3
IoT-based classroom monitoring	28	23.3

Note: "Regular use" reflects consistent weekly engagement; "High reliance" indicates frequent dependence on AI-generated recommendations.

18.3% a high dependency on IN-FACI activity. Use during professional development was very low (0.9 hours, SD. 8; 20.0% regular use). Overall, teachers used AI tools on average 10.4 hours per week (SD = 4.2), noting both the selective but also meaningful adoption of AI across instructional practices.

14.5.3 Emergent Dimensions of Teacher Identity

An exploratory factor analysis using the TIPS revealed that there were four factors of teacher identity in AI co-teaching, indicating a significant re-conceptualization of professional self-concept compared to traditional beliefs about teacher identity (see Table 14.3–14.6).

TABLE 14.3
Factor Analysis of Dimensions of Teacher Identity (N = 120)

Factor	Eigenvalue	% Variance	α	Sample Items	Factor Loadings
Factor 1: pedagogical curator	8.42	26.30%	0.89	"I see my role as selecting and adapting AI-generated materials"	0.78
				"My expertise lies in judging what works for my students"	0.82
				"I am a curator of learning resources, including AI-produced ones"	0.76
				"Evaluating instructional quality is more important than creating from scratch"	0.71
Factor 2: data interpreter	4.67	14.60%	0.86	"A key part of my job is making sense of analytics about student learning"	0.81
				"I translate data insights into actionable teaching strategies"	0.79
				"Understanding learning analytics is central to my professional identity"	0.74
				"I am skilled at reading patterns in student performance data"	0.77
Factor 3: ethical overseer	3.21	10.00%	0.84	"I am responsible for ensuring AI recommendations are fair and appropriate"	0.77
				"My role includes protecting students from potential harms of AI systems"	0.83
				"I must critically evaluate AI outputs rather than accepting them automatically"	0.72
				"Monitoring AI systems for bias is part of my professional responsibility"	0.69
Factor 4: relational anchor	2.89	9.00%	0.87	"My most important role is providing emotional support that AI cannot offer"	0.85
				"I am the human connection in an increasingly technological classroom"	0.8
				"Building relationships with students distinguishes me from AI"	0.78
				"My value as a teacher lies in understanding students as whole persons"	0.73
Total variance explained		59.90%			

TABLE 14.4
Regression Results Predicting Teacher Identity Dimensions (N = 120)

Factor	Eigenvalue	% Variance	α	Sample Items	Factor Loadings
Factor 1: pedagogical curator	8.42	26.30%	0.89	"I see my role as selecting and adapting AI-generated materials"	0.78
				"My expertise lies in judging what works for my students"	0.82
				"I am a curator of learning resources, including AI-produced ones"	0.76
				"Evaluating instructional quality is more important than creating from scratch"	0.71
Factor 2: data interpreter	4.67	14.60%	0.86	"A key part of my job is making sense of analytics about student learning"	0.81
				"I translate data insights into actionable teaching strategies"	0.79
				"Understanding learning analytics is central to my professional identity"	0.74
				"I am skilled at reading patterns in student performance data"	0.77
Factor 3: ethical overseer	3.21	10.00%	0.84	"I am responsible for ensuring AI recommendations are fair and appropriate"	0.77
				"My role includes protecting students from potential harms of AI systems"	0.83
				"I must critically evaluate AI outputs rather than accepting them automatically"	0.72
				"Monitoring AI systems for bias is part of my professional responsibility"	0.69
Factor 4: relational anchor	2.89	9.00%	0.87	"My most important role is providing emotional support that AI cannot offer"	0.85
				"I am the human connection in an increasingly technological classroom"	0.8
				"Building relationships with students distinguishes me from AI"	0.78
				"My value as a teacher lies in understanding students as whole persons"	0.73
Total variance explained		59.90%			

Note: β = standardized regression coefficient. *$p < .05$, *$p < .01$, $p < .001$.

The four-component solution accounted for 59.9% of the overall variance and had good internal consistency (all $\alpha > .84$) and support adequate assessment of separate identity dimensions. The Kaiser-Meyer-Olkin sampling adequacy index was .87, and the Bartlett's test of sphericity was significant ($\chi^2 = 2847.23$, $p < .001$), which supported the adequacy of factor analysis.

Factor 1: Pedagogical curator (26.3% of variance, $\alpha = .89$) was highlighted as the prevailing dimension of identity. Teachers who support this identity dimension perceive their expertise to be in the judicious use, adaptation, and personalization of resources rather

TABLE 14.5
Comparison of Perceived Authority Before vs. After AI Adoption (N = 120)

Authority Domain	Before AI M (SD)	After AI M (SD)	Difference	t	p	d
Knowledge transmission	4.32 (0.58)	3.47 (0.81)	−0.85	−9.87	<.001	−1.21
Curriculum and content selection	4.18 (0.64)	3.71 (0.76)	−0.47	−5.64	<.001	−0.67
Assessment and evaluation	4.41 (0.52)	3.62 (0.79)	−0.79	−9.23	<.001	−1.17
Classroom management	4.28 (0.61)	4.19 (0.68)	−0.09	−1.24	0.218	−0.14
Pedagogical decision-making	4.35 (0.56)	3.89 (0.72)	−0.46	−6.02	<.001	−0.71
Ethical judgment and student welfare	4.47 (0.51)	4.62 (0.49)	0.15	2.68	0.008	0.3
Relational and emotional support	4.39 (0.58)	4.71 (0.44)	0.32	5.21	<.001	0.62

Note: Authority rated on 5-point scale (1 = very low authority to 5 = very high authority). Positive difference indicates increased authority; negative indicates decreased authority.

than being generators themselves. This is a reconceptualization of pedagogical authority from origination to curation, from "I create all instructional materials" to "I make expert judgments about sources and the extent to which I need to transform and adapt them to the students in my community." The high proportion of the variance accounted for by this factor indicates that curation is a prominent organizing principle in teacher identity within AI co-teaching environments.

Factor 2: Data interpreter (14.6% of variance, $\alpha = .86$) on the rise of analytics literacy as a central identity category. Teachers placing themselves in higher part of this dimension have come to take on interpreting data as a core professional responsibility, and they are translating

TABLE 14.6
Qualitative Themes in Teacher Identity Transformation

Theme	Core Insight	Representative Quotations
1. Liberation and Loss	AI eases routine work but reduces creative ownership	"AI saves planning time, but I miss creating lessons from scratch. It feels like I'm losing part of my identity as a teacher." (Priya, 8 years, Private Secondary, Mathematics)
2. Negotiating Trust	Trust in AI is selective and experience-based	"Analytics help identify struggling students, but understanding *why* they struggle needs human judgment." (Anita, 6 years, Private Higher Education, English)
3. Visibility and Validation	Data-driven work gains visibility; emotional labor remains invisible	"Dashboards show administrators that I track progress, but emotional support never shows up." (Meera, 5 years, Private Secondary, Biology)
4. Algorithmic Accountability	Teachers feel monitored and pressured by AI systems	"When I don't follow AI suggestions, I feel the need to justify myself." (Lakshmi, 7 years, Government Secondary, Chemistry)
5. Redefining Expertise	Expertise shifts from content mastery to judgment and relationships	"My expertise is no longer knowing everything, but understanding students and context." (Kavita, 9 years, Private Secondary, English)
6. Generational Divide	Younger teachers adapt easily; veterans face identity disruption	"AI tools feel natural to me." (Neha, 3 years, Private Secondary, Computer Science) / "After 22 years as an expert, deferring to algorithms is difficult." (Ramesh, 22 years, Government Secondary, Social Studies)

between algorithmic outputs and pedagogical work. This identity factor reflects the ubiquity of learning analytics dashboards (72.5 percent of sample), and presents a new form of expertise discussed in teacher education discourse. Teachers who hold this identity do not view data analysis as an auxiliary activity, but as integral to one's professional competence.

Factor 3: Ethical overseer (10.0% of variance, $\alpha = .84$), demonstrates a teacher's sense of ownership regarding the monitoring and governing of AI systems in order to safeguard students' well-being. This theme includes healthy skepticism of algorithmic suggestions, concern for fairness and bias, and safeguarding against potential negative consequences. Educators who would claim this identity position themselves as required counterbalances to algorithmic authority, privileging human judgment for the "best interests" and "welfare" of students in AI systems. The domain of teachers as ethical arbitrators. This dimension indicates that teachers are asserting moral high ground in an area where human judgment is still paramount.

Factor 4: Relational anchor (9.0% of variance, Cronbach's $\alpha = .87$), attend to the affective, human dimensions of teaching that AI is unable to duplicate. High scores on this dimension reflect teachers' perception of themselves as valuable professionals based on emotional bonding, interpersonal help, and relationship building. This identity dimension could be understood as the defensive reaction to a technology's invasion; claims by teachers on relation work as their incommensurable professional domain. The focus on "the human connection" and support that "AI can't provide" implies the teachers are distinguishing themselves from AI because of certain aspects unique to what it means to be human.

The traditional identity dimensions, which are epistemic authority (acting as the sole source of knowledge) as well as autonomous decision maker (being in charge of one's own teaching), are conspicuously missing from this structure. These conventional dimensions d/f either loaded poorly onto factors (e.g., egocentric volatile/low certainty), or were distributed across multiple factors, indicating that AI co-teaching had comprehensively disturbed these identity bases. The rise of curator, interpreter, overseer, and anchor personas presages a re-imagining of the role of teacher in algorithmic-system enabled instruction.

14.5.4 Predictors of Identity Transformation

Multiple regression was used to investigate how AI use intensity, years of teaching experience, institutional context, and prior technology training predict endorsement of the four emerging issues. These analyses reveal what enables versus constrains identity transformation.

AI utilization intensity emerged as a strong and consistent predictor of change in teacher identity. In addition, teachers who used AI more frequently were significantly more likely to identify as data interpreters ($\beta = .51, p < .001$) and pedagogical curators ($\beta = .42, p < .001$). This implies that continued interaction with dashboards, analytics, and AI output require teachers to re-skill themselves not just as deliverers of content but also between the collection points for data and the entry points for learning versus progress logs. The Relational Anchor role even exhibited a small positive association with AI use ($\beta = .19, p < .05$), which may suggest that increased exposure to interactions with AI could heighten teachers' consciousness of the aspects of their work that are distinctively human.

Identity shifts were prompted in significantly different ways by teaching experience. Older teachers were less likely to be willing to become pedagogical curator ($\beta = -.18, p < .05$) and data interpreter ($\beta = -.23, p < .01$) roles, which indicates a more gradual integration to AI-powered workflows. Experience, on the other hand, acted to further increase identification with RG ($\beta = .31, p < .001$), indicating that senior teachers value more the students-relationship as a support to maintain them professionally updated.

Institutional and disciplinary contexts also played a role. Curator adopter innovation was also more likely to be adopted in the case of teachers who worked in private school settings ($\beta = .21$, $p < .05$) and STEM teachers were more likely to embrace data-oriented identities ($\beta = .24, p < .01$).

Previous training in technology or higher digital literacy also bolstered curator, data interpreter and ethical overseer roles with the most robust effect on data interpretation. The models as a whole accounted for a large amount of variance in AI-compatible identities, especially Data Interpreter ($R^2 = .47$), as well as pedagogical curator ($R^2 = .38$), and relational and ethical roles seemed to be influenced by broader personal and institutional factors than just the use of technology.

14.5.5 Authority Reconfiguration

The perceptions of the professional authority of teachers underwent a significant shift post-adoption with implications in different teaching domains. Such an analysis would suggest not that authority is simply being eroded everywhere, but rather that it is being redistributed in a more variegated way across various dimensions of the teaching work.

In all domains of teaching, teachers described their perceived authority as having decreased substantially due to the introduction of AI. The largest decline was in knowledge reporting (M difference = −0.85, $t = -9.87$, $p < .001$, $d = -1.21$) and examination/evaluation (M difference = −.79, $t = -9.23$, $p < .001$, $d = -1.17$). These very large effect sizes represent significant identity collapse in core teaching roles. With students being able to get AI-generated explanations and tutorial videos, alongside with instant answering when asking a question, teachers do not see themselves as the first or authoritative source. Likewise, teachers feel they have less control over evaluation when algorithms grade assignments, provide feedback, and recommend a level of mastery for students.

The data on the loss of power for teachers in regard to curriculum and content selection were unambiguous (M difference = −0.47, $t = -5.64$, $p <. 001$, $d = -0.67$), suggesting a move from complete creative control toward selection and refining of AI materials. An analogous pattern was observed in pedagogical decision making (M difference = −0.46, $t = -6.02$, $p < .001$, $d = -0.71$), indicating that pacing, differentiation, and intervention recommendations are largely driven by AI systems as opposed to individual teacher judgment.

At the same time, not all authority suffered a weakening. Differences in classroom management were largely maintained (M difference = −0.09, $p = .218$), which further reminds us that discipline and organization remain essentially human-run. Teachers, in particular, reported that they were more authority than moral judgment (M difference = +.15, $p = .008$) and significantly more so in relational and emotional support (M difference = +0.32, $p < .001$). In composite, these changes reveal a paradox; whereas teachers sense loss of control in cognitive and instructional matters they also perceive an organizing of authority in ethical and relational domains, thus indicating that it is more about a renegotiation than erosion of teacher position.

14.5.6 Lived Experiences: Qualitative Themes

Analysis of interviews and classroom observations suggested there were six key ways in which teachers experience and make sense of identity change. These themes offer significant explanations for how teachers make meaning of identity transformation, and negotiate new professional self-concept.

Theme 1: Liberation and Loss mirrors the profound ambivalence teachers hold for AI co-teaching. Most agreed that the automation of such mundane work was all well and good, but expressed a sense of loss at no longer having creative control. PriyaTh feeling of : Liberation and Loss mirrors the profound ambivalence teachers hold for AI co-teaching. Most agreed that the autoevidence for the principle that all too often, efficiency advances at the expense of authorship and professional pride. This tension was evident in 23 of the 30 interviews.

Theme 2: Negotiating Trust Teachers Develop Selective, Experience-Based Trust in AI. Anita believed the analytics for giving her early warning signs of students who were struggling (but not for why they struggled), and Suresh fell back on his instincts when the advice from an AI did not align with what he said or felt about how to proceed. Teachers had more confidence in AI for quantitative tasks than inferential or relational decisions. This was a theme in 27 interviews.

Theme 3: Visibility and Validation This is about how AI makes some work visible, at the same time obscuring other work. Meera had appreciated the dashboards that memorialized her monitoring work, but Deepak fretted that emotional labor abor to proceed. Teachers had more confidence in AI fteachers were feeling pressured to focus on measurable items. Nineteen of the interviews, particularly those conducted in government facilities, included this theme.

Theme 4: Algorithmic Accountability This theme reflects on increasing sentiments of surveillance. those conducted in government facilities, included this themeollow through on an A.I. recommendation, even though it was just implied, feels less right," Lakshmi and Arun wrote. Autonomy was subtly limited via data logging, as well as normative pressure. This concept was explicitly discussed in 18 interviews.

Theme 5: Rediscovering Expertise demonstrates teachers actively working to redefine identity, in this case, around human assets. Kavita accentuated judgment and student understanding rather than content mastery, and Vikram talked about skill in making sense of what the data do not present. This theme was present in 26 interviews, particularly among curatorial teachers and those adopting high ethical roles.

Theme 6: Generational Divide The generation gap is especially stark in terms of career stage. For Neha, AI seemed normal while Ramesh struggled to reconcile it with a longstanding identity as a classroom expert. This theme in 24 interviews is consistent with the quantitative data.

In general, the transformation of teacher identity was characterized as ambivalence rather than a binary option (accept or reject). Teachers had agency to redefine their roles, but the institutional context, training, and design of AI were crucial in determining whether or not AI was experienced as empowering or threatening.

14.6 DISCUSSION

14.6.1 Interpreting Identity Transformation

Taken together, the findings of both the quantitative and qualitative data suggest that AI co-teaching does not transform teacher identity in a single or homogeneous manner. Instead, it generates multiple but overlapping shifts in the ways that teachers perceive their professional identity. The four identity patterns identified—Pedagogical Curator, Data Interpreter, Ethical Overseer, and Relational Anchor—are indicative of a distinct departure from traditional understandings of the teacher as an all-knowing fount of knowledge who held complete pedagogic sovereignty. Instead of entirely "replacing" teacher identity, the AI situates it in new domains and responsibilities.

Drawing on role theory, AI introduces expectations that traditionally have not been a core of teaching, where teachers are expected to make sense of data (or lack thereof), monitor algorithms, and select among the artifacts produced by AI. But also de-emphasizes original content creation and independent judgment. This forces teachers to perform a new form of identity work that reconciles traditional professional values and technological demands. And while the stats suggest that for many, there is an orderly and planned path to making this transition, it's also clear that these journeys can be emotionally charged; a sense of loss and a real resistance to change are common themes.

The enormous priority attributed to that of the Pedagogical Curator points toward a significant recasting of what we understand pedagogic expertise to be. Expertise today is not in creating content; it's knowing how to judge quality, relevance, and fit with a student's context. It's not skill with a capital S that's lost, but rather a redefinition of what we mean by skill—and one that teachers themselves may find less visible or less rewarding.

Socio-technical That these identities are shifting reflects how we build and deploy AI today. The Data Interpreter comes as a reaction to the rise of analytics dashboards, and the Ethical Overseer from teachers wanting to override, contest or localize AI decisions. Finally, the Relational Anchor position reflects teachers' attempt to protect their humanity from technology's reconfiguration of the classroom.

14.6.2 The Authority Paradox

A key finding of the study is how teacher authority is unevenly reconfigured. Teachers perceive a clear loss of authority in terms of knowledge transmission and assessment, but also report gains in ethical or relational dimensions. This tendency in AI co-teaching suggests that "undermining" teacher authority is not simply about taking away from it (sub-trahere) but re-making it—changing how teacher authority "works," or more precisely, what it comes from.

Over time, the professional status of teachers was based on subject matter knowledge and having a way to attest that learning had occurred. These were jobs with class and institutional authority. This authority is easily challenged by AI systems that provide immediate answers, author material, and are auto-graded. The teachers are not the only source of knowledge or evaluation.

At the same time, teachers are asserting new forms of authority. That the rise of ethical authority is coupled with an increasing awareness that A.I. systems need human judgment to guard against bias, bad recommendations, and excessively narrow views about what learning means. Educators are also beginning to see their duties extend toward ensuring technology treats students fairly and equitably—a change from the University of Michigan to groups like Black Lives Matter at School Trace, in part, protest efforts. The increased relational authority is evidence of teachers appropriating emotional bonding, caring, and contextual knowledge as their core professional assets. These are areas where having a human being on the scene rather than a computer really does make a difference.

But this shift is also forcing us to confront some hard truths about what teaching will look like down the road. That kind of ethical and relational work is much needed, but it is also far less public and harder to quantify than academic success. A merit system which just rewards performance based on numbers may lead to teachers losing respect and independence. The long-term effects will depend on whether institutions and policymakers consider these human dimensions of teaching to be core as well as periphery to the quality of education.

14.6.3 Trust, Control, and Accountability Negotiations

The category of Negotiating Trust indicates that teachers do not simply accept or reject the AI. They are not transactional but conditional, they believe or trust (as per the context and experience). That's right—teachers trust AI for operationally competent functions such as calculations that allow for grading or when an algorithm can detect patterns in performance, but they are overwhelmingly skeptical of anything which requires interpretation, judgment or pedagogic sensibility. That speaks to an excellent understanding of where AI is applicable and where human skill remains a necessity.

Meanwhile the findings suggest an imbalance between trust and control. Making things even more complicated, teachers are supposed to follow AI's advice or explain why they did not, but remain accountable alone for the results of teaching. And it can create moral conflict, because teachers are on the hook for decisions being made in systems they might not have full control over or understanding of. The algorithmic provocateur of Algorithmic Accountability features writes in a similar vein how "data dashboards and activity logs monitor subtly, nudging actors to comply with algorithmic norms as it does so, even while teachers retain formal authority".

Taken together, these dynamics also suggest the need for more robust governance structures. Such models should explicitly reinforce teacher judgment as the primary, apportion responsibility equitably between teachers, institutions, and developers, defend teachers who override AI by effectively placing responsibility on the AI that was overruled where applicable, and mandate transparency that enables educators to confidently assess when AI does stupid things.

14.6.4 Convergences and Contradictions with Prior Literature

Our results are also consistent with and contrast with existing studies on AI in education. In line with previous studies, we find that teachers' work transitions from creation of content to curation and assessment), mixed emotions of use between empowerment and anxiety (Tsou et al., 2024), and the important role played by training and explainability in AI use. The evolution of teachers as curators and interpreters also aligns with a larger body of literature on emerging professional roles in technology-rich classrooms.

At the same time however, our findings qualify optimistic statements about AI-enabled augmentation. Although previous research on AI suggests that it enhances professional judgment by freeing educators from menial tasks, teachers' perceived authority in key functions (e.g., knowledge transfer, assessment, and pedagogical decision-making) is significantly eroded in our findings. These declines, which are accompanied by large effect sizes, signal more profound identity disturbance as opposed to typical role change. Qualitative themes of Liberation and Loss and Algorithmic Accountability also suggest that AI co-teaching creates the experience of constraint, surveillance, and creative autonomy loss as well as benefits such as time savings.

Finally, the contribution is novel in extending these dynamics in the context of India. Only 14.2% of teachers also said they had received broad formal training in the predominant use of ICT technology, Mr. Zur said, adding that the results revealed systemic failings in how AI was being implemented and taught across the Global South. In a domain where "teacher knows best" has been a longstanding theme, AI technologies challenging epistemic control could potentially lead to further professional divides. This means that the social context is important and should play a more explicit roles in future AI in education work.

14.7 IMPLICATIONS AND RECOMMENDATIONS

14.7.1 Implications for Teacher Practice

In AI co-taught scenarios, teachers can exercise professional agency by reframing the roles they play explicitly. The envelopment of such emergent roles as curate, interpret, shepherd and anchor should be counted genuine expertise instead as evidence of deskilling From the viewpoint of media systems. Curation and interpretation of data itself requires skill that is the product of a level judgment, contextual sensitivity and pedagogical cognition which appears to be uniquely human. It is thus crucial to develop a critical aliteracy. This involves knowing how AI systems function, recognizing their limitations and biases, comparing outputs with professional understanding, and overriding algorithmic suggestions as needed. Teachers could also narrate and valorize relational and ethical work that AI cannot perform by drawing on narratives, storytelling, qual data to make this work visible. Finally, collective identity work and professional learning communities and networks can provide support for teachers to share stories with peers and negotiate institutional expectations around the use of AI.

14.7.2 Implications for Institutions and Policymakers

If teacher agency is to be genuinely supported, however, institutions must work on continuously developing teachers' ability but also their willingness to develop resonant with identity just as much as skill mastery. Clear governance structures are needed to define who does what work when AI shapes instructional decision making, while protecting teachers as professionals. Transparence and explainability should be the standard in all such AI systems that potentially impact education, teachers must be able to understand all recommendations. Assessment systems have to protect against reductive data-driven accountability and must make explicit the relational, ethical, and supervisory aspects of teaching. Policies should also support participatory design practices that involve teachers in AI adoption and enactment decisions.

14.7.3 Implications for AI System Designers

AI tools should be designed with teacher-in-the-loop technologies where teachers are the ultimate decision makers. Clear context-sensitive rationales should be provided for recommendations. Systems should advance the full range of teaching effort rather than most, if not all, qualitative and relational features that guide the work that teachers do rather than minimize them in order to scale them up and include components for teacher learning and collaboration. Longitudinal impact assessments should consider how teachers' identity, authority, well-being and long-term professional trajectories are affected; not merely efficiency or learning outcomes.

14.8 LIMITATIONS AND FUTURE RESEARCH

There are a number of limitations of this study that further suggest valuable directions for future research. The cross-sectional nature of the study provides a snapshot in time of teacher identity within AI uptake. Teaching identities are developed over time and long-term studies should be conducted to determine whether the identity shifts found in these data consolidate, recede or shift as more experience is gained. Future work could investigate whether educators who refuse AI tools initially eventually become complacent users, or if early adopters feel concerned about loss of expertise or professional freedom.

Second, the authority essay was partly based on teachers' thoughts about their pre-AI positions and might be subject to memory or current perspectives. The frustrations or advantages teachers currently face as a result of AI use might also influence the way they remember past authority. Even better would be to have studies that take initial measurements prior to the adoption of AI, and then measure at some point after it.

Third, although the sample included diversity in subject area and institutional type, it was geographically constrained. Comparative work across regions and countries would serve to disentangle how cultural perceptions of teacher authority and variations in institutional support influence identity change.

Finally, future investigations need to go beyond surveys and investigate classroom interactions, student perceptions, and institutional contexts in order to present a more nuanced view on the ways AI is remixing teaching in everyday practice.

14.9 CONCLUSION

This study explored how AI co-teaching tools (re)configure teacher professional identity so that teachers can shift from becoming the sole arbiter of instruction to forging decisions together, as a "unit of two," with algorithmic systems. Centered on a theory-based mixed-methods design and based on 120 teachers across secondary and tertiary education in India, the study reveals distinct changes in the manner that teachers perceive their authority, autonomy, and professional roles.

The findings indicate that the co-teaching siloed AI model doesn't yield merely a substitution, nor an easy amplification of teaching labor. Instead, it's creating a re-orientation of professional lines. Through these four identity dimensions—Pedagogical Curator, Data Interpreter, Ethical Overseer, and Relational Anchor, traditional notions of power are challenged away from a standard understanding of teacher power. Teachers report having far less control over the transmission of knowledge and assessment, and far more power in ethical judgment and relational support. This shift doesn't mean an erosion of expertise as much as a relocation of the "where" and "how" of professional value.

Identity transformation is uneven. The qualitative results imply that participants experience ambivalent feelings, with a feeling of empoweredness and efficiency, but also the death of autonomy, trust with AI related to transparency and type of task.

These findings have far-reaching implications in teacher training, academic administration as well as AI system design. In the development of teachers, identity change (becoming a better version of oneself) and critical AI literacy should be made explicit. "These tools are a significant step in the deprofessionalizing of teaching," he added, and yet they "do not make decisions about drug-purchasing or recidivism without human input." Institutions need governance structures that protect teachers' judgment, respect the relational, ethical, and transparent nature of this work. Designers need to be able to build teacher-in-the-loop systems that do not diminish professional agency, but actually support it.

Ultimately, AI integration isn't just a technical shift; it's an overhaul of teaching itself. Whether this becomes a corrosive or rejuvenating force will perhaps depend on how identity, authority, and human judgment come to matter for the design and governance of educational AI.

REFERENCES

Chu, T., Jiang, A. L., & Wu, N. (2025). Negotiating identity in co-teaching discourse: power dynamics among L1 and L2 English-speaking teachers in EFL contexts. *SAGE Open*, *15*(4). https://doi.org/10.1177/21582440251390318

Hermansen, H. (2017). Knowledge relations and epistemic infrastructures as mediators of teachers' collective autonomy. *Teaching and Teacher Education*, *65*, 1–9. https://doi.org/10.1016/j.tate.2017.03.003

Kim, J. (2024). Types of teacher-AI collaboration in K-12 classroom instruction: Chinese teachers' perspective. *Education and Information Technologies*, *29*(13), 17433–17465. https://doi.org/10.1007/s10639-024-12523-3

Kim, K., & Kwon, K. (2025). From co-design to co-teaching: a comprehensive approach to an integrated AI curriculum in middle school STEM education. *Smart Learning Environments*, *12*(1), 57. https://doi.org/10.1186/s40561-025-00413-1

Kwet, M., & Prinsloo, P. (2020). The 'smart' classroom: a new frontier in the age of the smart university. *Teaching in Higher Education*, *25*(4), 510–526. https://doi.org/10.1080/13562517.2020.1734922

Memarian, B., & Doleck, T. (2024). Human-in-the-loop in artificial intelligence in education: a review and entity-relationship (ER) analysis. *Computers in Human Behavior: Artificial Humans*, *2*(1), 100053. https://doi.org/10.1016/j.chbah.2024.100053

Montobbio, F., Staccioli, J., Virgillito, M. E., & Vivarelli, M. (2024). The empirics of technology, employment and occupations: lessons learned and challenges ahead. *Journal of Economic Surveys*, *38*(5), 1622–1655. https://doi.org/10.1111/joes.12601

Nazaretsky, T., Ariely, M., Cukurova, M., & Alexandron, G. (2022). Teachers' trust in AI-powered educational technology and a professional development program to improve it. *British Journal of Educational Technology*, *53*(4), 914–931. https://doi.org/10.1111/bjet.13232

Nopas, D-siri. (2025). Algorithmic learning or learner autonomy? Rethinking AI's role in digital education. *Qualitative Research Journal*. https://doi.org/10.1108/QRJ-11-2024-0282

Ojetunde, S. M., & Ramnarain, U. (2025). Like objects or like subjects? Effects of student–robot interaction (SRI) and mathematical ability on students learning outcomes. *Smart Learning Environments*, *12*(1), 9. https://doi.org/10.1186/s40561-024-00345-2

Possaghi, I., Vesin, B., Zhang, F., Sharma, K., Knudsen, C., Bjørkum, H., & Papavlasopoulou, S. (2025). Integrating multi-modal learning analytics dashboard in K-12 education: insights for enhancing orchestration and teacher decision-making. *Smart Learning Environments*, *12*(1), 53. https://doi.org/10.1186/s40561-025-00410-4

Raviv, A., Bar-Tal, D., Raviv, A., Biran, B., & Sela, Z. (2003). Teachers' epistemic authority: perceptions of students and teachers. *Social Psychology of Education*, *6*(1), 17–42. https://doi.org/10.1023/A:1021724727505

Roll, I., & Wylie, R. (2016). Evolution and revolution in artificial intelligence in education. *International Journal of Artificial Intelligence in Education*, *26*(2), 582–599. https://doi.org/10.1007/s40593-016-0110-3

Schiff, D. (2020). Out of the laboratory and into the classroom: the future of artificial intelligence in education. *AI & Society*, *36*(1), 331–348. https://doi.org/10.1007/s00146-020-01033-8

Seufert, S., Sonderegger, S., & Hartmann, P. (2025). *From Tool to Teaching Partner: Integrating AI into Teacher Education with Intelligent-TPACK*, pp. 57–74. https://doi.org/10.1007/978-3-032-05817-1_3

Sony, M., & Naik, S. (2020). Industry 4.0 integration with socio-technical systems theory: a systematic review and proposed theoretical model. *Technology in Society*, *61*, 101248. https://doi.org/10.1016/j.techsoc.2020.101248

Spina, N. (2020). *Data Culture and the Organisation of Teachers' Work. Data Culture and the Organisation of Teachers' Work.* https://doi.org/10.4324/9780429056192

Teasley, S. D. (2019). Learning analytics: where information science and the learning sciences meet. *Information and Learning Sciences*, *120*(1–2), 59–73. https://doi.org/10.1108/ILS-06-2018-0045

Topali, P., Ortega-Arranz, A., Rodríguez-Triana, M. J., Er, E., Khalil, M., & Akçapınar, G. (2025). Designing human-centered learning analytics and artificial intelligence in education solutions: a systematic literature review. *Behaviour and Information Technology*, *44*(5), 1071–1098. https://doi.org/10.1080/0144929X.2024.2345295

Tsou, W., Lin, A. M. Y., & Chen, F. (2024). Co-journeying with ChatGPT in tertiary education: identity transformation of EMI teachers in Taiwan. *Language, Culture and Curriculum*, *37*(4), 529–543. https://doi.org/10.1080/07908318.2024.2362326

Zembylas, M. (2003). Emotions and teacher identity: a poststructural perspective. *Teachers and Teaching: Theory and Practice*, *9*(3), 213–238. https://doi.org/10.1080/13540600309378

Index

Adaptive learning systems 2, 8, 16
Administrative transformation 2, 9
AI ethics 11–14, 16
AI governance 11–14
AI literacy 16
AI-driven assessment 8–11
AI-driven education 1–18
AI-enabled classrooms 8–10
AI-enabled learning 125–139
AI-powered analytics 1–2
AI-ready schools 61–70
Algorithmic bias 4–5, 8–11
Artificial intelligence (AI) 1–18, 21–36, 53–70, 71–88, 117–139, 140–149, 161–180
    adoption barriers 3–5
    curriculum integration 15–16
    ethics 11–14
    governance 11–14
    policy frameworks 12
    resistance to change 3–5
    smart education ecosystems 2, 7–8
    sustainability concerns 140–149
Assessment systems 3, 8, 10, 125–139
Automated grading 3, 8
Automation in education 1–2

B

Behavioural analytics 8–10
Big data in education 2, 11
Blended learning 21–36
British universities 99–116

C

Change management 1–18
Chatbots 21–36, 53–60
Classroom automation 1–2
Cloud computing 37–52
Co-teaching tools 161–180
Collaborative learning 21–36
Compliance frameworks 11–14
Computational thinking 1–2
Connectivity infrastructure 4–5, 16
Continuous professional development (CPD) 13–14
Cybersecurity 4, 12–14

D

Data analytics 1–2, 8, 13
Data governance 4–5, 8–14
Data privacy 4–5, 8–12
Decision-making systems 2–4
Digital divide 5, 12, 16
Digital equity 5, 12
Digital infrastructure 1–2, 4–5, 13–16
Digital literacy 1–2, 13, 16
Digital pedagogy 13
Digital transformation 1–2, 12–16
Distance learning 21–36

E

e-learning ecosystems 21–36
EdTech 1–18, 89–98
Educational governance 1–2, 11–14
Educational innovation 1–18
Educational policy 11–16
Emerging economies 16
Environmental sustainability 140–149
Ethical AI 11–14
Ethics in education technology 4–5, 11–14

F

Faculty development 13–14
Feedback systems 53–60
Fourth Industrial Revolution 1
Future of education 1–18, 161–180

G

GenAI 99–116
Governance models 1, 11–14

H

Higher education 71–88, 125–139
Human–AI collaboration 161–180

I

Inclusive education 12, 16
Innovation ecosystems 1–18
Institutional readiness 13
Instructional innovation 53–60
Intelligent tutoring systems 1–2
Internet of Things (IoT) 1–18, 37–52, 71–88, 140–149
    campus automation 2
    classroom integration 37–52
    governance challenges 11–14
    infrastructure requirements 4–5
    sustainability issues 140–149

L

Learning analytics 2, 8, 16
Learning management systems (LMS) 21–36

M

Machine learning 2–4
Mobile learning 21–36

P

Pedagogical innovation 21–36, 53–60, 125–139
Personalized learning 1–2, 8, 117–124
Policy-to-practice transition 71–88
Predictive analytics 1–2, 8
Professional identity 3, 161–180
Prompt literacy 53–60

R

Regulatory compliance 4, 11–14
Resistance to change 1–18
Responsible AI 11–14

S

Smart campuses 2, 16
Smart classrooms 37–52, 71–88
Smart diagnostics 117–124
Social inequality 5, 12
Stakeholder engagement 8–10, 13–14
Student monitoring systems 8–11
Sustainability 12, 140–149
Sustainable digital transformation 12–16

T

Teacher identity transformation 161–180
Teacher training 2, 13–16
Technological change 1–18
Technology acceptance models 6–7
Technology integration 1–18
Trust in AI systems 8–10

U

Universal access 12

V

Virtual learning environments (VLEs) 21–36

W

Workforce readiness 1–2

For Product Safety Concerns and Information please contact our EU
representative GPSR@taylorandfrancis.com
Taylor & Francis Verlag GmbH, Kaufingerstraße 24, 80331 München, Germany

www.ingramcontent.com/pod-product-compliance
Ingram Content Group UK Ltd.
Pitfield, Milton Keynes, MK11 3LW, UK
UKHW052221190926
475702UK00006B/1145
9781041290612